Practical Digital Electronics

Pierre Pelloso

Conservatoire National des Arts et Métiers, Inqénieur du Professeur Enseignant à l' technique des Ecoles Nationales Supérieures des Arts et Métiers, I.U.T. de Créteil et à l'ESIGELEC de Rouen
France

Translated from the original French by
John C.C. Nelson
(Department of Electrical and Electronic Engineering, University of Leeds, UK)

A Wiley–Interscience Publication

JOHN WILEY & SONS
Chichester · New York · Brisbane · Toronto · Singapore

First published under the title *Pratique de l'électronique numérique* by Bordas
Dunod Gautuer-Villaes, Paris, 1983

Library of Congress Cataloging in Publication Data

Pelloso, Pierre.
Practical digital electronics.

Translation of: Pratique de l'électronique numérique.
'A Wiley–Interscience publication.'
Bibliography: p. 215
Includes index.
1. Digital electronics. I. Title.
TK7868.D5P4513 1985 621.3815 85-15524

ISBN 0 471 90733 2

British Library Cataloguing in Publication Data:

Pelloso, Pierre
Practical digital electronics.
1. Digital integrated circuits.
I. Title
621.381'73 TK7874

ISBN 0 471 90733 2

Printed and bound in Great Britain

Contents

Summary

Preface

Preface

Some readers may be surprised to see a selection of problems with solutions in a collection of technical works. In fact, the examples chosen are extracts from the everyday practice of digital technology. They make use of basic concepts, methods of solution, and current technology.

We have not been able, within the limits of this work, to cover complex applications; we have preferred to develop several problems restricted to the areas of basic functions, technology, and digital systems in general, including binary arithmetic. These problems have been given to many students of IUT and engineering colleges, in addition to industrial technicians and engineers in further education.

Each chapter begins with a brief reminder of the fundamental theory. The problems come next, followed by the solutions. Certain examples are presented in a manner which allows the reader a precise interpretation in one particular direction. Others, by contrast, leave complete freedom of interpretation and hence of the solution of the problem.

We would like to thank the publishers and in particular M. J. D. Bertron for the care which they have devoted to the realization of this work, Melle L. Renoux who has drawn the very numerous figures which illustrate our examples, and Mmes S. Sallier and J. Declercq who typed the manuscript.

1 Basic Functions

1.1 SUMMARY OF THEORY

A summary of binary algebra, limited to the essentials, is presented below.

Binary variable

x_i takes the values, or states, in $\{0, 1\}$

Boolean function of n binary variables

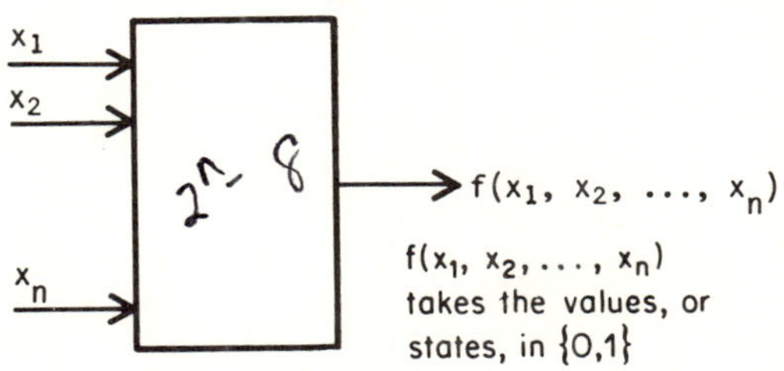

Figure 1.1

Truth table

This completely defines the function.

For n variables, it requires 2^n lines which represent all the combinations of these n variables.

There are $2^{(2^n)}$ different functions of n variables.

For $n = 1$

x	f_1	f_2	f_3	f_4
0	0	0	1	1
1	0	1	0	1

$f_2(x) = x$

$f_3(x) = \bar{x}$

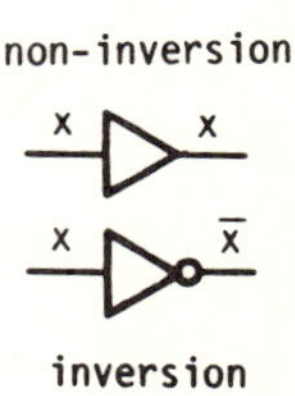

Figure 1.2

For $n = 2$

Complementary functions

x	y	f_1	f_2	f_3	f_4	f_5	f_6	f_7	f_8	f_9	f_{10}	f_{11}	f_{12}	f_{13}	f_{14}	f_{15}	f_{16}
0	0	0	0	0	0	0	0	0	0	1	1	1	1	1	1	1	1
0	1	0	0	0	0	1	1	1	1	0	0	0	0	1	1	1	1
1	0	0	0	1	1	0	0	1	1	0	0	1	1	0	0	1	1
1	1	0	1	0	1	0	1	0	1	0	1	0	1	0	1	0	1

Fundamental functions

*Complement, or inverse (see Figure 1.2)
*Sum (OR):

$$f_8(x, y) = x + y$$

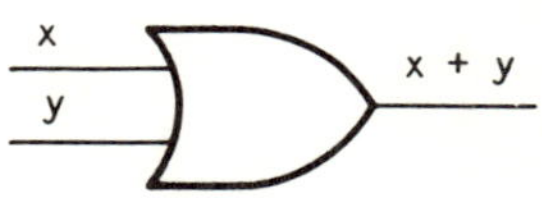

Figure 1.3

*Product (AND):

$$f_2(x, y) = x \cdot y$$

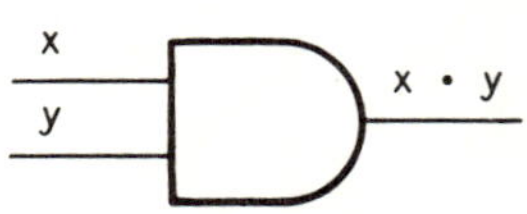

Figure 1.4

Properties of the fundamental functions

Involution	$\overline{(\bar{x})} = x$	
Commutativity	$x \cdot y = y \cdot x$	$x + y = y + x$
Associativity	$(x \cdot y) \cdot z = x \cdot (y \cdot z)$	$(x + y) + z = x + (y + z)$
Neutrality	$x \cdot 1 = 1 \cdot x = x$	$x + 0 = 0 + x = x$
Absorption	$x \cdot 0 = 0 \cdot x = 0$	$x + 1 = 1 + x = 1$
Idempotence	$x \cdot x = x$	$x + x = x$
Complementation	$x \cdot \bar{x} = 0$	$x + \bar{x} = 1$

Distributivity
$$\begin{cases} x \cdot (y + z) = (x \cdot y) + x \cdot z \\ x + y \cdot z = (x + y) \cdot (x + z) \end{cases}$$

de Morgan's laws $\begin{cases} \overline{x \cdot y} = \bar{x} + \bar{y} \\ \overline{x + y} = \bar{x} \cdot \bar{y} \end{cases}$

Inclusive laws

$$x + x \cdot y = x \qquad x \cdot (x + y) = x$$

$$x + \bar{x} \cdot y = x + y \qquad x \cdot (\bar{x} + y) = x \cdot y$$

$$x \cdot y + \bar{x} \cdot z + y \cdot z = x \cdot y + \bar{x} \cdot z \qquad (x + y) \cdot (\bar{x} + z) \cdot (y + z) = (x + y) \cdot (\bar{x} + z)$$

In general:

$$[(x_1 \cdot x_2) \cdot x_3 \cdot \ldots] \cdot x_n = x_1 \cdot x_2 \cdot \ldots \cdot x_n \qquad [(x_1 + x_2) + x_3) + \ldots] + x_n = x_1 + x_2 + \ldots + x_n$$

takes the 1 state if all the variables are at 1 — takes the 0 state if all the variables are at 0

All functions can be written in the form of *Shannon's theorems*:

$$f(x_1, \ldots, x_p, \ldots, x_n) = \bar{x}_p \cdot f(x_1, \ldots, 0, \ldots, x_n) + x_p \cdot f(x_1, \ldots, 1, \ldots, x_n)$$

$$f(x_1, \ldots, x_p, \ldots, x_n) = [x_p + f(x_1, \ldots, 0, \ldots, x_n)] \cdot [\overline{x_p} + f(x_1, \ldots, 1, \ldots, x_n)]$$

These two theorems allow all functions to be written in the form:

- — of a sum of products ($\sum\prod$) called the *first canonical form*;
- — of a product of sums ($\prod\sum$) called the *second canonical form*.

NOR and NAND functions

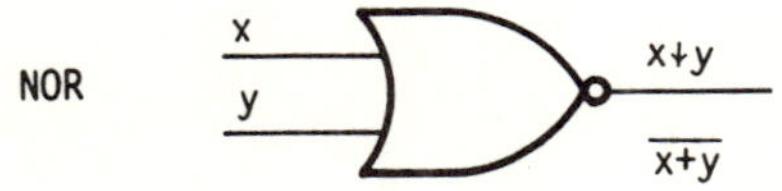

Figure 1.5

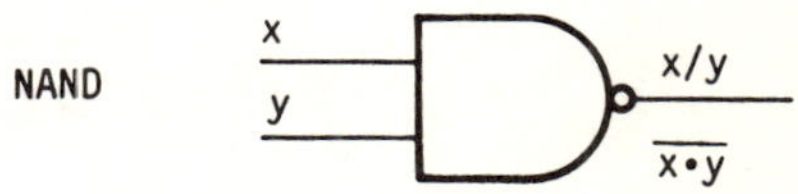

Figure 1.6

$$f_9(x, y) = x \downarrow y = \overline{x + y} = \bar{x} \cdot \bar{y} \qquad f_{15}(x, y) = x/y = \overline{x \cdot y} = \bar{x} + \bar{y}$$

Properties:

commutative $x \downarrow y = y \downarrow x$ — commutative $x/y = y/x$

associative $(x \downarrow y) \downarrow z \neq x \downarrow (y \downarrow z)$ — associative $(x/y)/z \neq x/(y/z)$

neither neutral nor absorbent — neither neutral nor absorbent

Particular cases:

$x \downarrow 0 = \bar{x}$ inversion — $x/0 = 1$

$x \downarrow 1 = 0$ | $x/1 = \bar{x}$ inversion

$x \downarrow x = \bar{x}$ inversion | $x/x = \bar{x}$ inversion

$x \downarrow \bar{x} = 0$ | $x/\bar{x} = 1$

$(x \downarrow y) \downarrow 0 = x + y$ sum | $(x/1)/(y/1) = x + y$ sum

$(x \downarrow 0) \downarrow (y \downarrow 0) = x \cdot y$ product | $(x/y)/1 = x \cdot y$ product

hence the complete functional operation

$(a + b) \cdot (c + d) \cdot \ldots (y + z)$
$= (a \downarrow b) \downarrow (c \downarrow d) \downarrow \ldots \downarrow (y \downarrow z)$

hence the complete functional operation

$(a \cdot b) + (c \cdot d) + \ldots + (y \cdot z)$
$= (a/b)/(c/d)/\ldots/(y/z)$

In general:

$x_1 \downarrow x_2 \downarrow \ldots \downarrow x_n = \overline{x_1 + x_2 + \ldots + x_n}$
therefore

$(a + b + \ldots + k) \cdot (m + n + \ldots + x)$
$= (a \downarrow b \downarrow \ldots \downarrow k) \downarrow (m \downarrow n \downarrow \ldots \downarrow x)$

$(x + y + \ldots + t) \downarrow (a + b + \ldots + h)$
$= x \downarrow y \downarrow \ldots \downarrow t \downarrow a \downarrow b \downarrow \ldots \downarrow h$

$x \downarrow y \downarrow z = (\overline{x \downarrow y}) \downarrow z = x \downarrow (\overline{y \downarrow z})$

$x_1/x_2/\ldots/x_n = \overline{x_1 \cdot x_2 \cdot \ldots \cdot x_n}$
therefore

$(a \cdot b \cdot \ldots \cdot k) + (m \cdot n \cdot \ldots \cdot x)$
$= (a/b/\ldots/k)/(m/n/\ldots/x)$

$(x \cdot y \cdot \ldots \cdot t)/(a \cdot b \cdot \ldots \cdot h)$
$= x/y/\ldots/t/a/b\ldots/h$

$x/y/z = (\overline{x/y})/z = x/(\overline{y/z})$

Exclusive OR function

(XOR, modulo 2 sum)

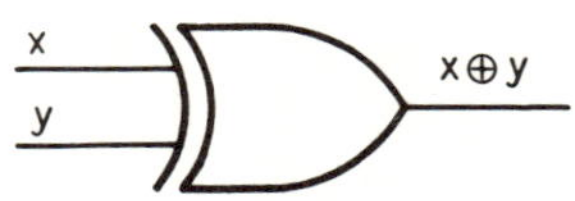

Figure 1.7

$$z = f_7(x, y) = x \oplus y = (\bar{x} \cdot y) + (x \cdot \bar{y}) = (x + y) \cdot (\bar{x} + \bar{y}) = (x + y) \cdot \overline{x \cdot y}$$
$$= [x/(x/y)]/[y/(x/y)] = \{[x \downarrow (x \downarrow y)] \downarrow [y \downarrow (x \downarrow y)]\} \downarrow 0$$

Properties:

commutative	$x \oplus y = y \oplus x$
associative	$(x \oplus y) \oplus z = x \oplus (y \oplus z)$
neutral	$x \oplus 0 = x$
not absorbent	

Particular cases:

$x \oplus 0 = x$	$x \oplus x = 0$
$x \oplus 1 = \bar{x}$	$x \oplus \bar{x} = 1$
$x \oplus \bar{y} = \overline{x \oplus y}$	$\bar{x} \oplus \bar{y} = x \oplus y$

$c = a \oplus b \Rightarrow a = c \oplus b$ or $b = a \oplus c$

In general:

$$x_1 \oplus x_2 \ldots \oplus x_n$$

takes the 1 states if an odd number of variables is at 1, and takes the 0 state if an even number of variables is at 1.

$$\{[(x_1 \oplus x_2) \oplus x_3] x_4 \oplus \ldots\} \oplus x_n = x_1 \oplus x_2 \oplus x_3 \oplus \ldots \oplus x_n$$

$$x \oplus y \oplus \bar{z} = \overline{x \oplus y \oplus z}$$

$$x \oplus \bar{y} \oplus \bar{z} = x \oplus y \oplus z$$

COINCIDENCE function

Figure 1.8

$$z = f_{10}(x, y) = x \odot y = \overline{x \oplus y} = (x \cdot y) + (\bar{x} \cdot \bar{y}) = (\bar{x} + y) \cdot (x + \bar{y}) = (xy) + \overline{(x + y)}$$
$$= [x \downarrow (x \downarrow y)] \downarrow [y \downarrow (x \downarrow y)] = \{[x/(x/y)]/[y/(x/y)]\}/1$$

Properties:

commutative	$x \odot y = y \odot x$
associative	$(x \odot y) \odot z = x \odot (y \odot z)$
neutral	$x \odot 1 = x$
not absorbent	

Particular cases:

$x \odot 0 = \bar{x}$	$x \odot x = 1$
$x \odot 1 = x$	$x \odot \bar{x} = 0$
$x \odot \bar{y} = x \oplus y$	$\bar{x} \odot \bar{y} = x \odot y$

In general:

$$x_1 \odot x_2 \odot \ldots \odot x_n$$

takes the 1 state when all the variables are in the 0 state or the 1 state.

Comments:

$$x \odot y = \overline{x \oplus y}$$

$$(x \odot y) \odot z = x \oplus y \oplus z$$

$$[(x \odot y) \odot z] \odot t = \overline{x \oplus y \oplus z \oplus t}$$

etc.

1.2 STATEMENT OF EXERCISES AND PROBLEMS

The following exercises and problems are intended to illustrate the preceding summary.

1.E.1 Numbering

1. Consider variables which can take *two* distinct states represented by the symbols □ and △.

1a. For one, two, and three variables, how many combinations are possible? Draw the corresponding tables.

1b. Generalize to n variables.

2. Repeat the previous questions assuming variables taking *three* distinct states represented by the symbols □, △, and ○.

3. Generalize to n variables taking q distinct states.

4. How many functions can be created from n variables taking *two* distinct states?

5. Repeat the previous question generalized to n variables taking q distinct states.

1.E.2 Shannon's relations

1. Show that a function of one variable can be written:

$$f(x) = \bar{x} \cdot f(0) + x \cdot f(1).$$

2. Hence deduce that a function of two variables can be written:

$$f(x, y) = \bar{x} \cdot \bar{y} \cdot f(0, 0) + \bar{x} \cdot y \cdot f(0, 1) + x \cdot \bar{y} \cdot f(1, 0) + x \cdot y \cdot f(1, 1).$$

3. Generalize to n variables.

4. Hence deduce that:

$$\begin{aligned} f(x_{n-1}, \ldots, x_{k+1}, x_k, x_{k-1}, \ldots, x_0) = {} & \overline{x_k} \cdot f(x_{n-1}, \ldots, x_{k+1}, 0, x_{k-1}, \ldots, x_0) \\ & + x_k \cdot f(x_{n-1}, \ldots, x_{k+1}, 1, x_{k-1}, \ldots, x_0). \end{aligned}$$

This relation expresses Shannon's fisrt expansion theorem.

5. *Application*: Simplify the relation $f(x, y, z) = x \cdot y + \bar{x} \cdot z + y \cdot z$.

1.E.3 The 2-input multiplexer, a fundamental device

The 2-input multiplexer is defined by the symbol of Figure 1.9 and by the following properties:

$$\text{while} \begin{cases} A = 0, & S = E_0 \\ A = 1, & S = E_1 \end{cases}$$

A is called the *address input*, S the *output*, and E_1, E_0 the *principal inputs*.

1. Draw a Karnaugh map (K map) which represents this function.

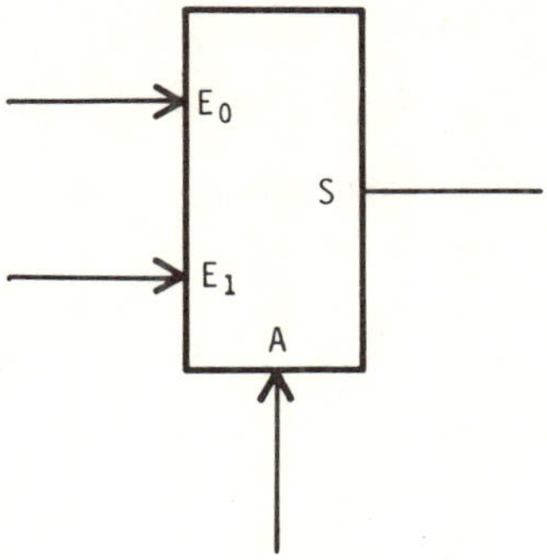

Figure 1.9

2. Deduce the simplified expression $S = f(A, E_1, E_0)$.
3. Syntheszie this operator:
3a. using sum, product and inverting operations;
3b. using NAND operations;
3c. using NOR operations.
4. Show that it is possible to realize all the functions of a single variable by using a single multiplexer in each case.
5. Show that it is also possible to realize all the functions of two variables by using one or more multiplexers: this operator is therefore functionally complete. Realize the functions $x_1 + x_0$; $x_1 \cdot x_0$; $\overline{x_1 + x_0}$; $\overline{x_1 \cdot x_0}$; $x_1 \oplus x_0$.
6. Show that the 2-input multiplexer realizes Shannon's first expansion theorem: it is therefore a fundamental device. It can be shown that it is possible to generate any function of n variables. Treat the general case of functions of two and three variables.
7. *Application*: Treat the case of Figure 1.10.

dc \ ba	00	01	11	10
00	1	0	0	1
01	0	0	0	0
11	0	0	1	1
10	1	0	0	0

Figure 1.10

1.3 SOLUTIONS

1.S.1 Numbering

1a. A variable x_0 allows two combinations to be defined: these correspond to each state taken separately.

x_0
□
Δ

* Adding the variable x_1: two combinations of x_0 can be associated with each state of x_1, hence four combinations.

x_1	x_0
□	□
□	Δ
Δ	□
Δ	Δ

* Adding the variable x_2: four combinations of x_1 and x_0 can be associated with each state of x_2, hence eight combinations.

x_2	x_1	x_0
□	□	□
□	□	Δ
□	Δ	□
□	Δ	Δ
Δ	□	□
Δ	□	Δ
Δ	Δ	□
Δ	Δ	Δ

1b. It is clear that each additional variable doubles the number of combinations C. Therefore n binary variables can define $C = 2^n$ combinations.

2. One ternary variable allows three combinations to be defined.

x_0
□
Δ
○

* One additional ternary variable triples the number of combinations, hence nine combinations.

x_1	x_0
□	□
□	Δ
□	○
Δ	□
Δ	Δ
Δ	○
○	□
○	Δ
○	○

3. Let p be the number of combinations associated with n variables taking q distinct states. One more variable allows $C = q \times p$ combinations.

Now for $n=1$ there are $C=q$ combinations,
for $n=2$ there are $C=q\times q=q^2$ combinations,
⋮
and for $n=n$ there are $C=q^n$ combinations.

Example: Let $n=2$ and $q=10$ states represented by the figures 0, 1,..., 9. The well known table is obtained which defines $10^2=100$ combinations.

x_1	x_0
0	0
0	1
⋮	⋮
0	9
1	0
⋮	⋮
1	9
⋮	⋮
9	0
⋮	⋮
9	9

4. The 2^n combinations of n variables can be put, as we have already seen, in the form of a table of 2^n lines.

2^n lines

x_{n-1}	$\cdots$	x_0	$f_0(\ )$	$\cdots$	$f_{F-1}(\ )$

A function f_i is obtained in association with each one of these lines regardless of the □ state or the △ state. In making this association in all possible ways, all the functions which form a finite set of n elements are obtained.

It is clear that this applies to the preceding problem, since C lines allows the creation of 2^C combinations and hence functions. Now $C=2^n$, hence: $F=2^n$.

Example: $n=4$ binary variables allows $C=2^4=16$ combinations and $F=2^{16}=65\,536$ functions to be obtained.

5. Identical reasoning allows $F=q^{(q^n)}$ functions to be obtained.

Example: $n=2$ variables taking $q=10$ distinct states allow $C=10^2=100$ combinations and $F=10^{100}$ functions to be obtained.

Comment: The states □, △, ○,... can advantageously be replaced by 0, 1, 2,..., if this facilitates later mathematical treatment.

1.S.2 Shannon's relations

1. The relation:

$$f(x) = \bar{x} \cdot f(0) + x \cdot f(1) \tag{1}$$

must be verified for each value of x. Since x can take only two states represented by 0 and 1:

— for $x = 0$, expression (1) becomes:

$$f(0) = 1 \cdot f(0) + 0 \cdot f(1) = f(0);$$

— for $x = 1$, expression (1) becomes:

$$f(1) = 0 \cdot f(0) + 1 \cdot f(1) = f(1).$$

Expression (1) is therefore verified in all cases.

2. Using Shannon's theorem, the preceding demonstration allows us to write:

$$f(x, y) = \bar{x} \cdot f(0, y) + x \cdot f(1, y).$$

$$\begin{aligned}
\text{Similarly:}\ f(0, y) &= \bar{y} \cdot f(0, 0) + \bar{y} \cdot f(0, 1) \\
\text{and}\ f(1, y) &= \bar{y} \cdot f(1, 0) + y \cdot f(1, 1) \\
\text{Therefore:}\ f(x, y) &= \bar{x} \cdot [\bar{y} \cdot f(0, 0) + y \cdot f(0, 1)] \\
&\quad + x \cdot [\bar{y} \cdot f(1, 0) + y \cdot f(1, 1)] \\
&= (\bar{x} \cdot \bar{y} \cdot f(0, 0) + \bar{x} \cdot y \cdot f(0, 1) \\
&\quad + x \cdot \bar{y} \cdot f(1, 0) \\
&\quad + x \cdot y \cdot f(1, 1).
\end{aligned}$$

3. Since $f(x_{n-1}, \ldots, x_1, x_0)$

$$\begin{aligned}
&= \overline{x_{n-1}} \cdot \ldots \cdot \overline{x_1} \cdot \overline{x_0} \cdot f(0, \ldots, 0, 0) + \overline{x_{n-1}} \cdot \ldots \cdot \overline{x_1} \cdot x_0 \cdot f(0, \ldots, 0, 1) \\
&\quad + \ldots + x_{n-1} \cdot \ldots \cdot x_1 \cdot x_0 \cdot f(1, \ldots, 1, 1).
\end{aligned}$$

For one extra variable we have:

$$\begin{aligned}
&f(x_n, x_{n-1}, \ldots, x_1, x_0) \\
&\quad = \overline{x_n} \cdot f(0, x_{n-1}, \ldots, x_1, x_0) + x_n \cdot f(1, x_{n-1}, \ldots, x_1, x_0).
\end{aligned}$$

Now $f(0, x_{n-1}, \ldots, x_1, x_0)$

$$\begin{aligned}
&= \overline{x_{n-1}} \cdot \ldots \cdot \overline{x_1} \cdot x_0 \cdot f(0, 0, \ldots, 0, 0) + \overline{x_{n-1}} \cdot \ldots \cdot \overline{x_1} \cdot x_0 \cdot f(0, 0, \ldots, 0, 1) \\
&\quad + \ldots + x_{n-1} \cdot \ldots \cdot x_1 \cdot x_0 \cdot f(0, 1, \ldots, 1, 1)
\end{aligned}$$

and $f(1, x_{n-1}, \ldots, x_1, x_0)$

$$\begin{aligned}
&= \overline{x_{n-1}} \cdot \ldots \cdot \bar{x}_1 \cdot \bar{x}_0 \cdot f(1, 0, 0, \ldots, 0, 0) + \overline{x_{n-1}} \cdot \ldots \cdot \bar{x}_1 \cdot x_0 \cdot f(1, 0, \ldots, 0, 1) \\
&\quad + \ldots + x_{n-1} \cdot \ldots \cdot x_1 \cdot x_0 \cdot f(1, 1, \ldots, 1, 1).
\end{aligned}$$

From which $f(x_n, x_{n-1}, \dots, x_1\, x_0)$

$$= \overline{x_n} \cdot \overline{x_{n-1}} \cdot \dots \cdot \overline{x_1} \cdot \overline{x_0} \cdot f(0,0,..0,0) + \overline{x_n} \cdot \overline{x_{n-1}} \cdot \dots \cdot \overline{x_1} \cdot x_0 \cdot f(0,0,\dots 0,1)$$
$$+ \dots + x_n \cdot x_{n-1} \cdot \dots \cdot x_1 \cdot x_0 \cdot f(1,1,\dots,1,1).$$

4. The order in which the variables are treated is unimportant, and it is easy to show that:

$$f(x_{n-1}, \dots, x_{k+1}, x_k, x_{k-1} \dots, x_0)$$
$$= \overline{x_k} \cdot f(x_{n-1}, \dots, x_{k+1}, 0, x_{k-1}, \dots, x_0)$$
$$+ x_k \cdot f(x_{n-1}, \dots, x_{k+1}, 1, x_{k-1}, \dots, x_0).$$

x_k is said to be "put in the expansion".

5. Putting x in the expansion:

$$f(x, y, z) = \bar{x} \cdot f(0, y, z) + x \cdot f(1, y, z), \qquad (2)$$

with $f(0, y, z) = z + y \cdot z$ and $f(1, y, z) = y + y \cdot z$.

Putting z in the expansion of $(0, y, z)$:

$$f(0, y, z) = \bar{z} \cdot f(0, y, 0) + z \cdot f(0, y, 1),$$
hence $f(0, y, z) = \bar{z} \cdot (0) + z \cdot (1) = z$.

Similarly, putting y in the expansion of $f(1, y, z)$:

$$f(1, y, z) = \bar{y} \cdot f(1, 0, z) + z \cdot f(1, 1, z),$$
hence $f(1, y, z) = \bar{y} \cdot (0) + y \cdot (1) = y$.

Substituting in (2), we obtain the well known relation:

$$f(x, y, z) = x \cdot y + \bar{x} \cdot z + y \cdot z = x \cdot y + \bar{x} \cdot z.$$

1.S.3 The 2-input multiplexer

1. The K map is given in Figure 1.11.

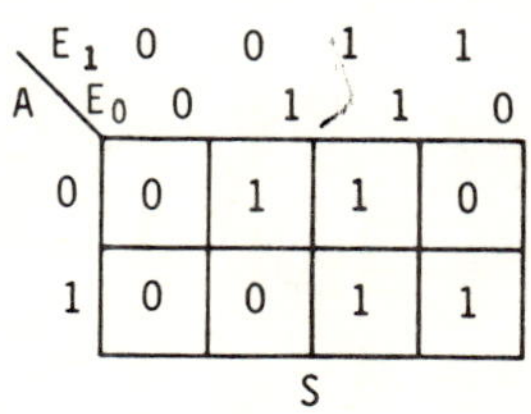

Figure 1.11

2. The expression are:

$$S = \bar{A} \cdot E_0 + A \cdot E_1 \qquad (1)$$
$$S = (\bar{A} + E_1) \cdot (A + E_0) \qquad (2)$$

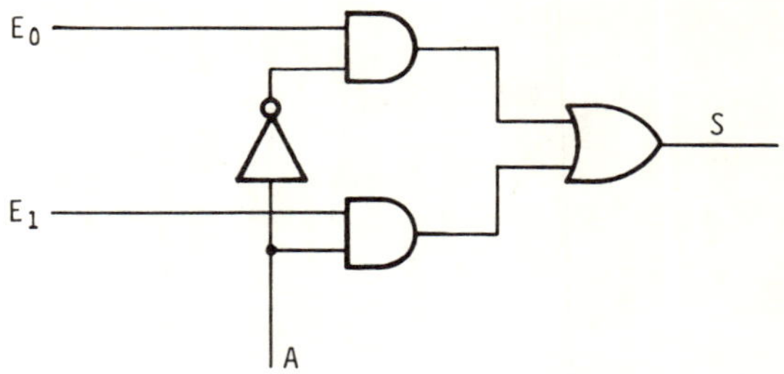

Figure 1.12

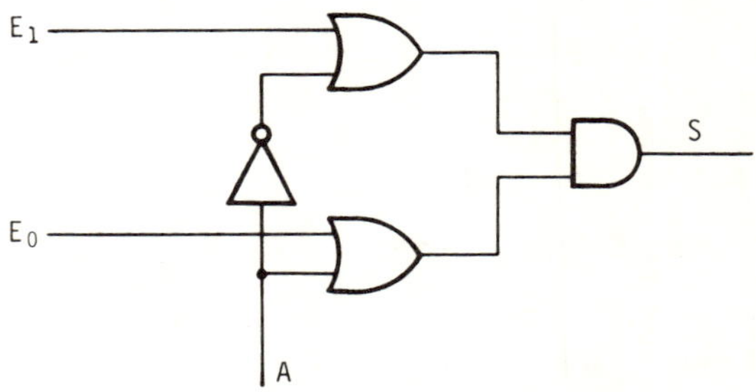

Figure 1.13

3a. The diagrams of Figures 1.12 and 1.13 correspond to the above expressions.

3b. The previous expression (1) can be written:

$$S = [(A/1)/E_0]/(A/E_1)$$

Figure 1.14 corresponds to this expression.

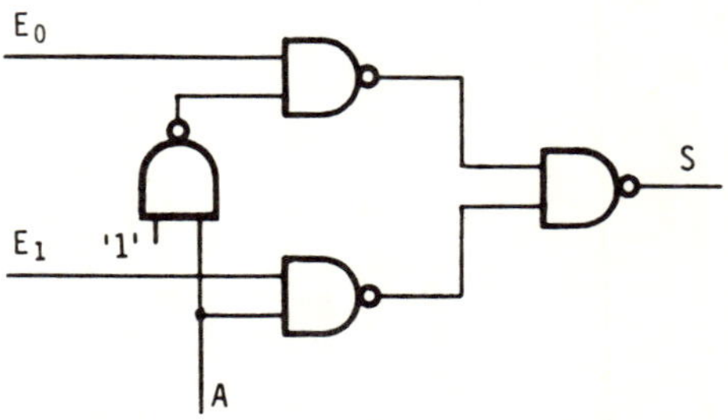

Figure 1.14

3c. Similarly expression (2) can be written:

$$S = [(A\downarrow 0)]\downarrow E_1\downarrow(A\downarrow E_0)$$

Figure 1.15 corresponds to this expression.

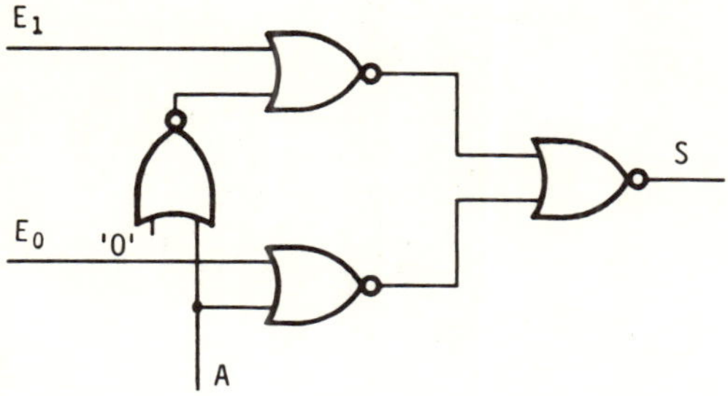

Figure 1.15

4. Compare Figure 1.11 and the table of functions of a single variable on page 1. Putting the variable x on the input A, and the coded combinations of the columns of Figure 1.11 on the inputs E_0 and E_1, we obtain Figure 1.16.

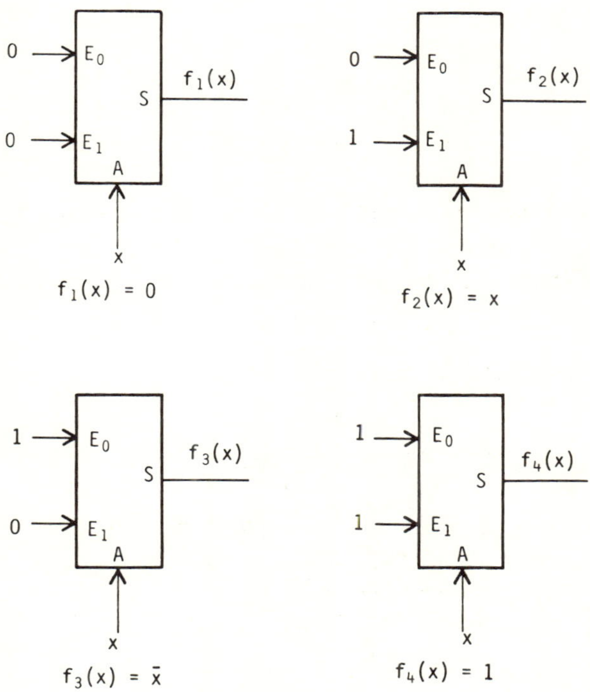

Figure 1.16

Only function f_3 is of interest.

5. Consider the K map of Figure 1.17, which represents all the functions of two variables.

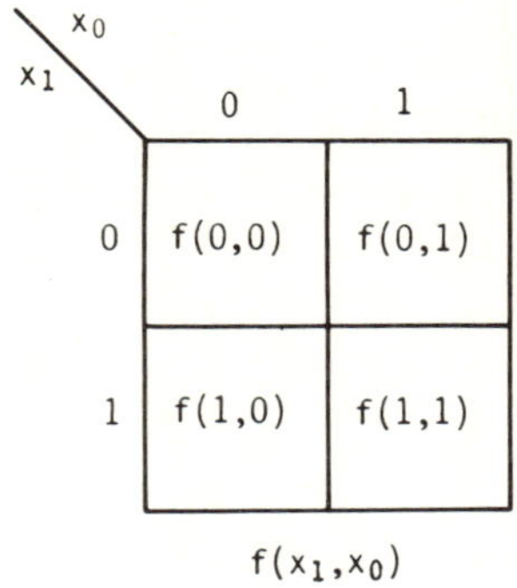

Figure 1.17

Consider also the multiplexer address table, given in Figure 1.18.

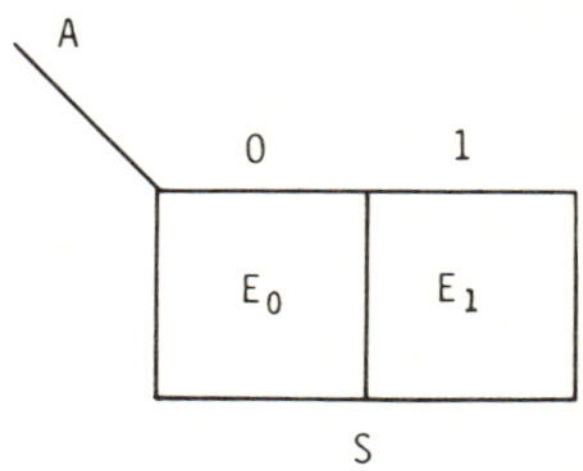

Figure 1.18

This table shows that if $A = 0$, $S = E_0$, and if $A = 1$, $S = E_1$.

Compare Figures 1.17 and 1.18. Let x_0 be associated with A; Figure 1.19 shows this assignment.

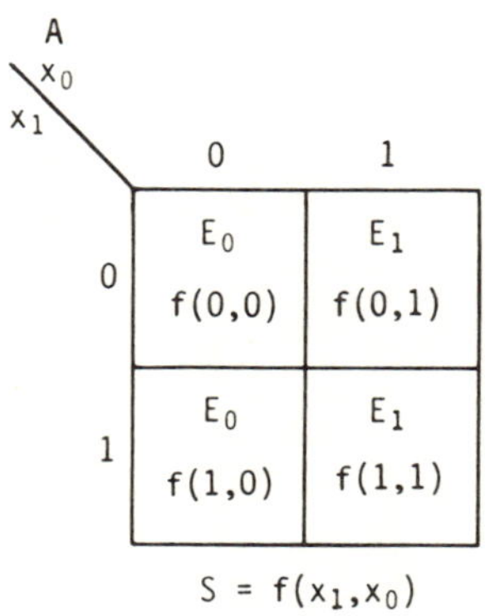

Figure 1.19

When $x_0 = 0$, f becomes a function of the variable x_1 only: $f(x_1, 0)$, and $S = E_0$. Therefore $S = E_0 = f(x_1, 0)$.

Figure 1.19 shows that there are four cases, summarized in Figure 1.20 according to the states of $f(0,0)$ and $f(1,0)$.

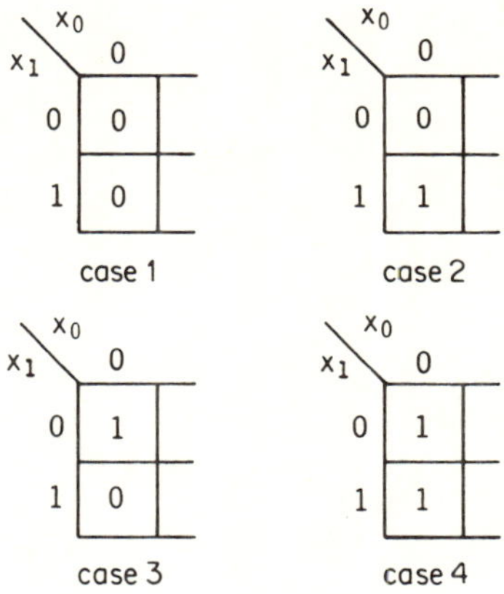

Figure 1.20

Case 1: For both values of x_1, $f(x_1, 0) = E_0 = 0$;
it is sufficient to put 0 on E_0.

Case 2: When $x_1 = 0$, $f(x_1, 0) = E_0 = 0$
When $x_1 = 1$, $f(x_1, 0) = E_0 = 0$;
it is sufficient to put x_1 on E_0.

Case 3: When $x_1 = 0$, $f(x_1, 0) = E_0 = 1$
When $x_1 = 1$, $f(x_1, 0) = E_0 = 0$;
it is sufficient to put $\bar{x}_1$ on E_0.

Case 4: For both values of x_1, $f(x_1, 0) = E_0 = 1$;
it is sufficient to put 1 on E_0.

The same reasoning applies when $x_0 = 1$, using the second column in Figure 1.19; in this case $S = E_1 = f(x_1, 1)$.

Figure 1.21 shows the required examples.

Comment: Another solution is obtained by putting x_1 on A and repeating the argument of Figure 1.19, as shown in Figure 1.22.

6. Shannon's expansion theorem allows us to write:

$$f(x_{n-1}, \ldots, x_{k+1}, x_k, x_{k-1}, \ldots, x_0)$$
$$= \bar{x}_k \cdot f(x_{n-1}, \ldots, x_{k+1}, 0, x_{k-1}, \ldots, x_0) + x_k \cdot f(x_{n-1}, \ldots, x_{k+1}, 1, x_{k-1}, \ldots, x_0) \quad (1)$$

The 2-input multiplexer realizes the function:

$$S = \bar{A} \cdot E_0 + A \cdot E_1 \quad (2)$$

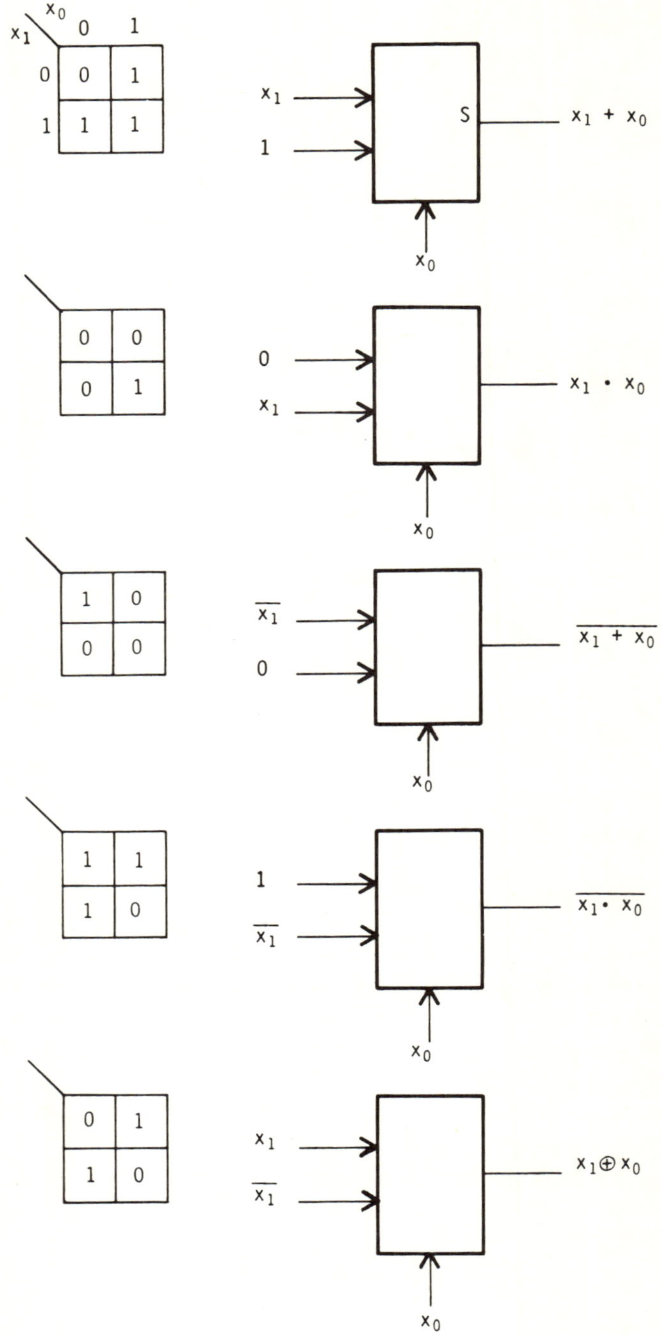
x0
x1
0
1
0
1
0
1
1
1
x1
1
S
x1 + x0
x0
0
0
0
1
0
x1
x1 · x0
x0
1
0
0
0
x1
0
x1 + x0
x0
1
1
1
0
1
x1
x1 · x0
x0
0
1
1
0
x1
x1
x1 ⊕ x0
x0

Figure 1.21

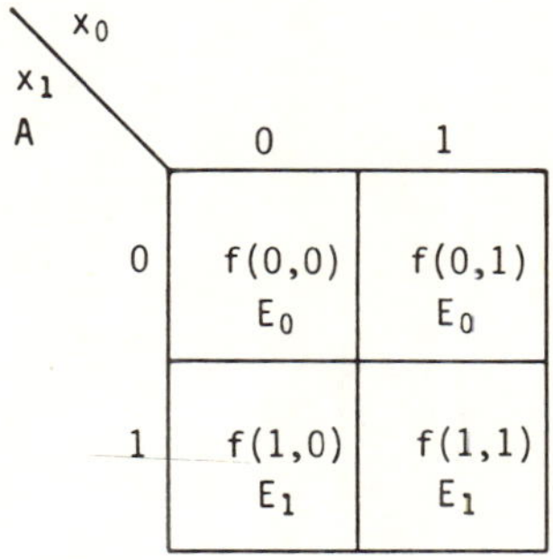

Figure 1.22

Equations (1) and (2) are implemented by Figure 1.23.

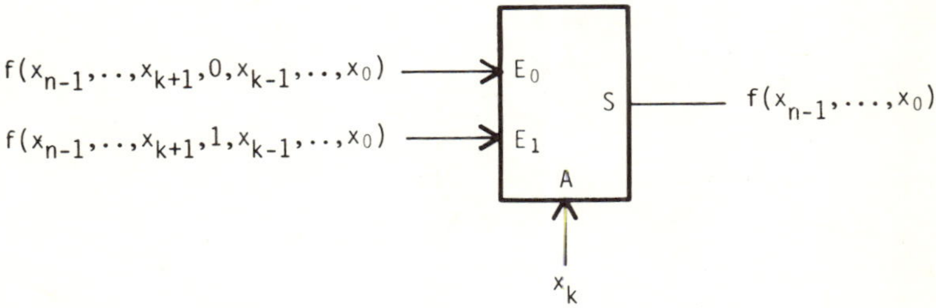

Figure 1.23

It is sufficient to use the same technique for all the variables in order to synthesize any function. It should be noted that there are n levels of multiplexers counting from right to left. The last level receives, on its inputs, the states taken by the function for each combination of the variables. The number of multiplexers in each level is twice the number in the previous level. In total there are s multiplexers where:

$$s = 1 + 2 + 4 + \ldots + 2^{n-1}.$$

This sum is that of a geometric progression of n terms, of which the first term $u = 1$ and, of course, $q = 2$, hence:

$$s = \frac{u(q^n - 1)}{q - 1} = \frac{1(2^n - 1)}{2 - 1} = 2^n - 1.$$

However, there can be simplifications.

Figure 1.24 treats the general case of functions of two variables, and Figure 1.25 that of functions of three variables.

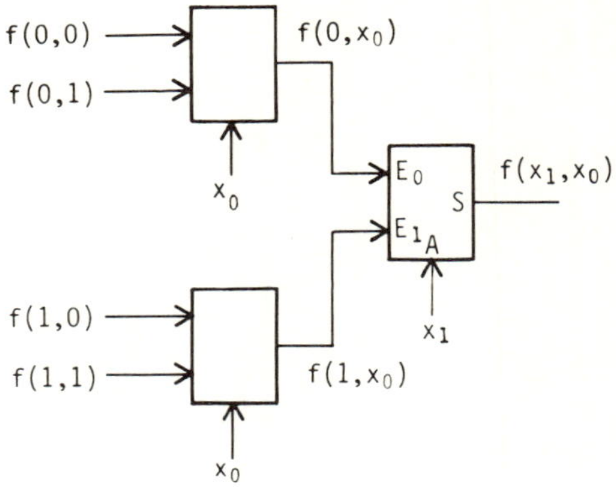

Figure 1.24

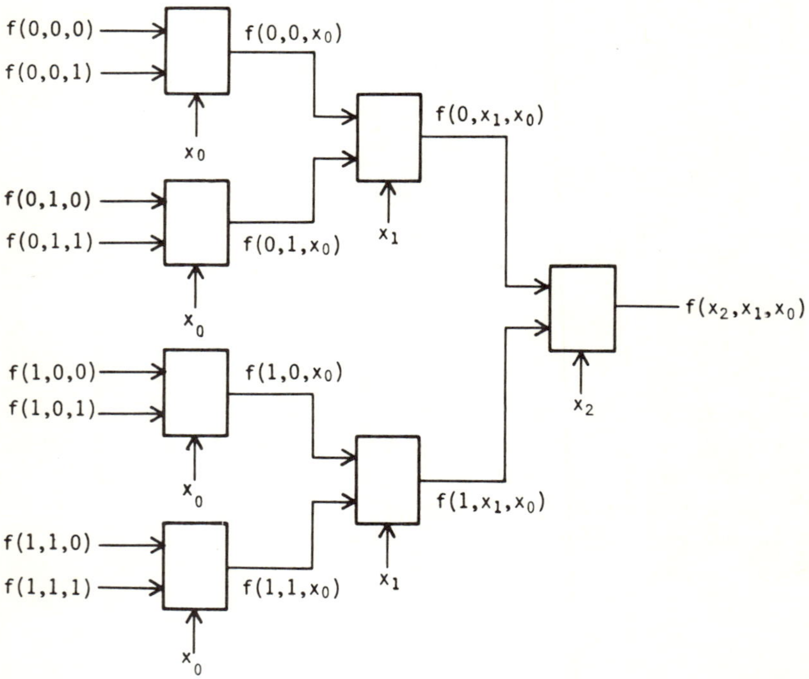

Figure 1.25

7. Figure 1.26 treats the proposed application.

The states of $f(d, c, b, a)$ are taken from the K map of Figure 1.10. Some simplifications have been made. The last level consists of a single multiplexer which inverts a.

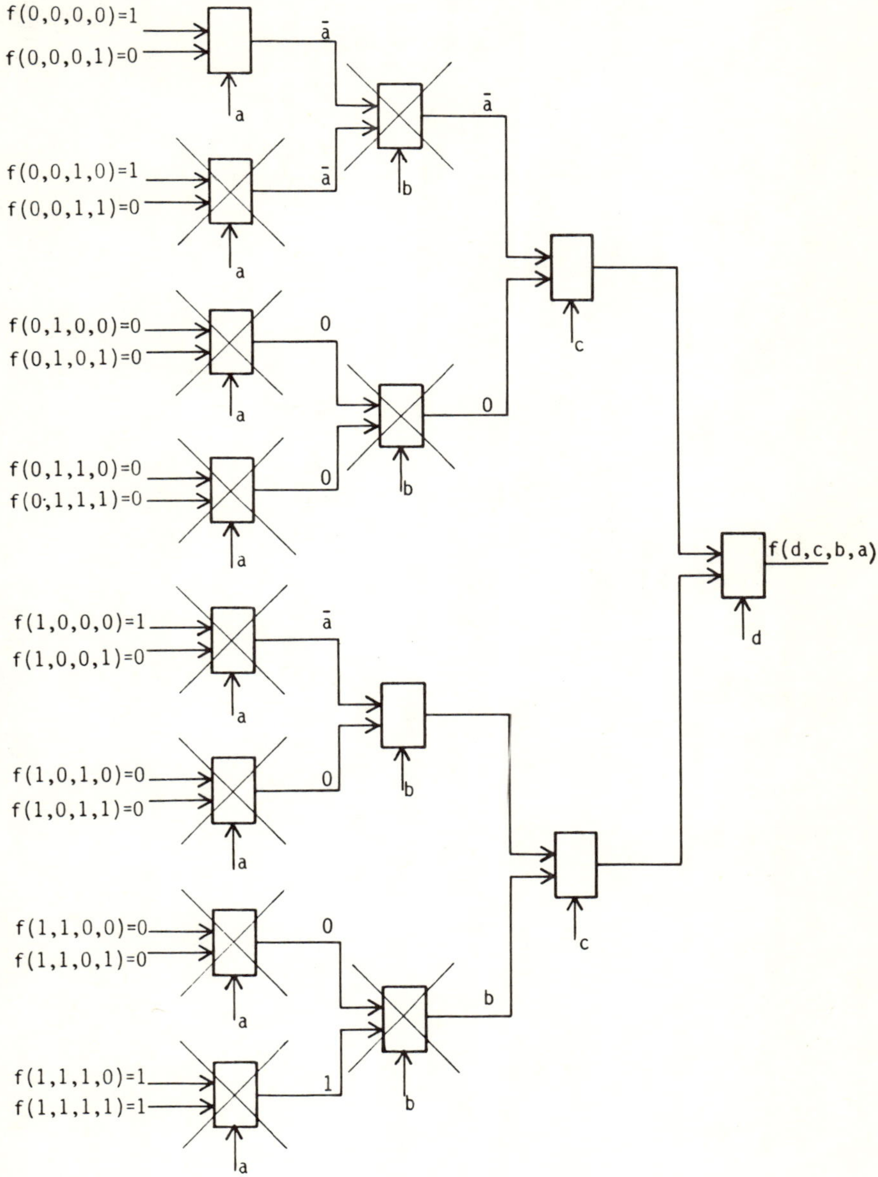

f(0,0,0,0)=1
f(0,0,0,1)=0
f(0,0,1,0)=1
f(0,0,1,1)=0
f(0,1,0,0)=0
f(0,1,0,1)=0
f(0,1,1,0)=0
f(0,1,1,1)=0
f(1,0,0,0)=1
f(1,0,0,1)=0
f(1,0,1,0)=0
f(1,0,1,1)=0
f(1,1,0,0)=0
f(1,1,0,1)=0
f(1,1,1,0)=1
f(1,1,1,1)=1
a
b
c
d
ā
0
1
f(d,c,b,a)

Figure 1.26

Finally, Figure 1.27 is obtained, which shows that:

$$f(d,c,b,a) = \bar{d}\bar{c}\bar{a} + dcb + d\bar{c}\bar{b}\bar{a}$$
$$= dcb + \bar{c}\bar{a}\cdot(\bar{d} + d\bar{b}),$$

since $\bar{d} + d\bar{b} = \bar{d} + \bar{b}$:

$$f(d,c,b,a) = dcb + \bar{d}\bar{c}\bar{a} + \bar{c}\bar{b}\bar{a},$$

which is the relation given by the K map of Figure 1.10.

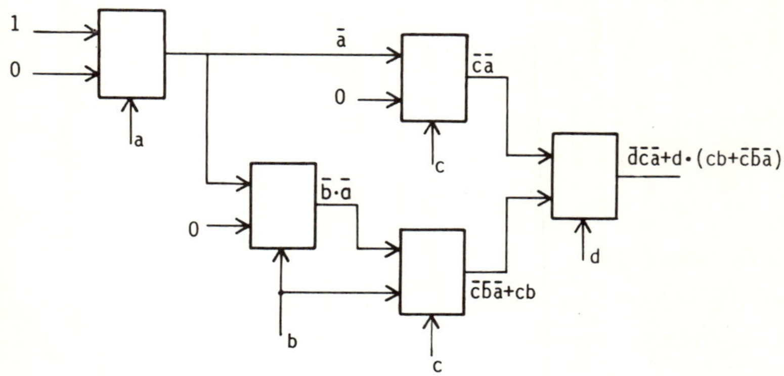

Figure 1.27

2 *Technology of Integrated Circuits*

2.1 SUMMARY OF THEORY

2.1.1 TTL technology

The basic circuit of TTL technology is given in Figure 2.1. In positive logic this arrangement realizes the function $S = \overline{E_1 \cdot E_2}$

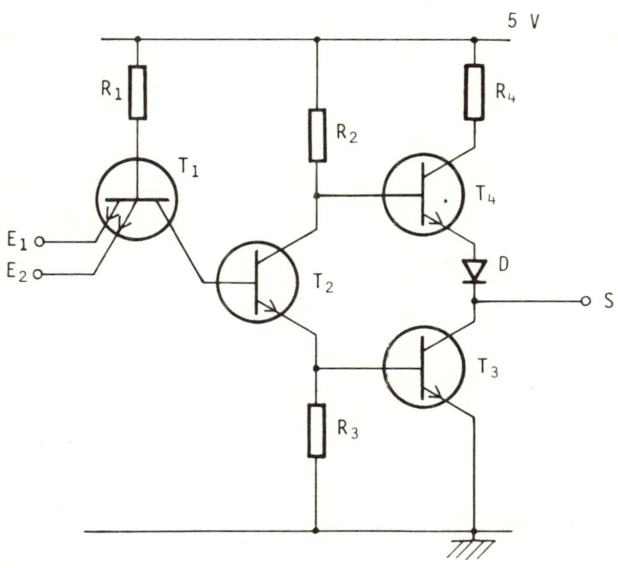

Figure 2.1

The output stage, consisting of T_3 and T_4, is called a 'totem-pole'.

Two cases, summarized in Figures 2.2 and 2.3, can occur. A third case represented by Figure 2.4, can be anticipated. It is possible only if the circuit is modified.

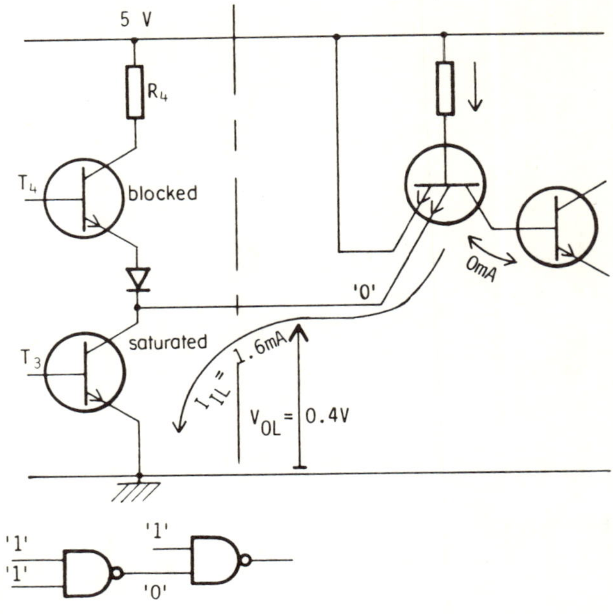

Figure 2.2

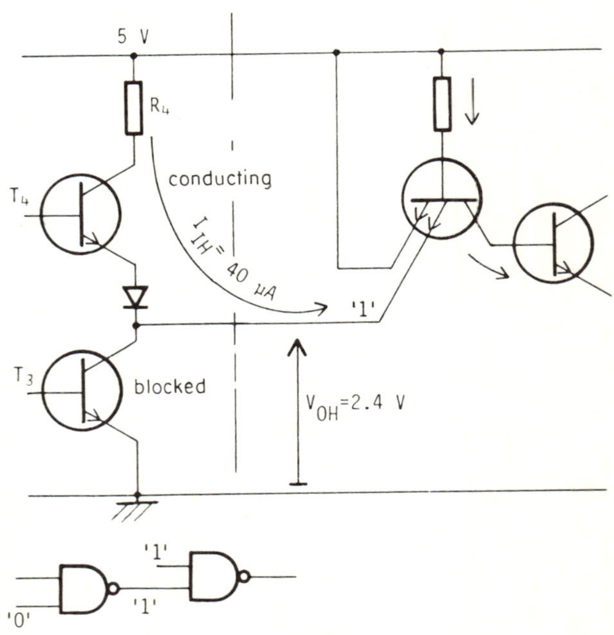

Figure 2.3

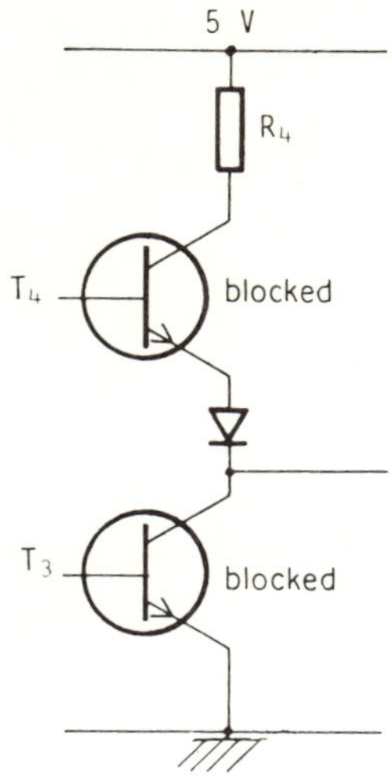

Figure 2.4

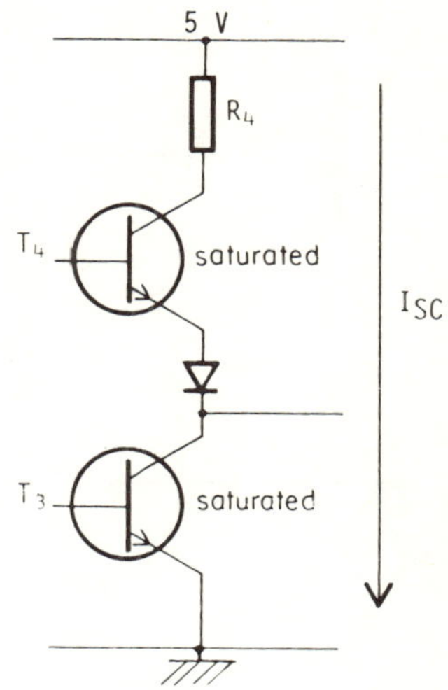

Figure 2.5

A fourth case, represented by Figure 2.5, is not permissible since the components R_4, T_4, D, and T_3 are not rated to carry I_{SC}. However, it can be encountered when circuits are interconnected, as will be shown.

I_{IH} and I_{IL} (see Figures 2.2 and 2.3) define the unit of current called a *TTL* load. All currents, with certain exceptions, are multiples of these quantities.

2.1.2 CMOS technology

The basic circuit of the CMOS family is the inverter given in Figure 2.6.

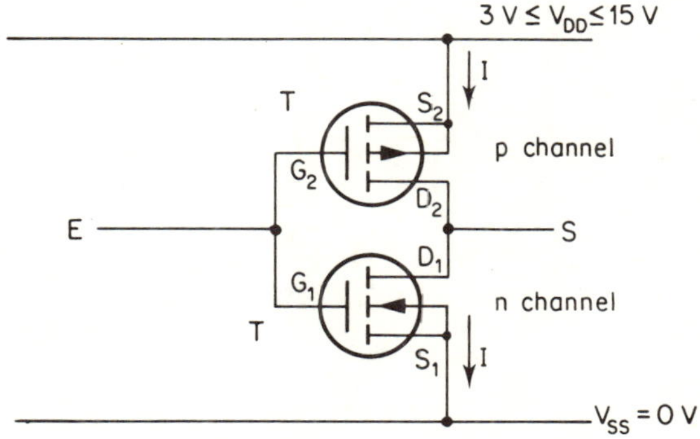

Figure 2.6

The operation can be summarized by the two graphs given in Figures 2.7 and 2.8

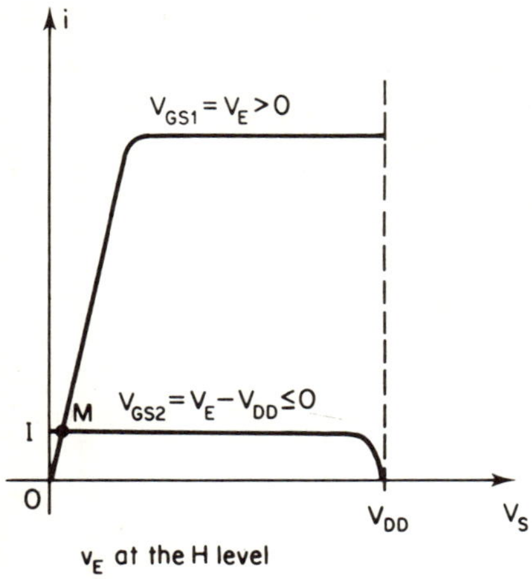

Figure 2.7

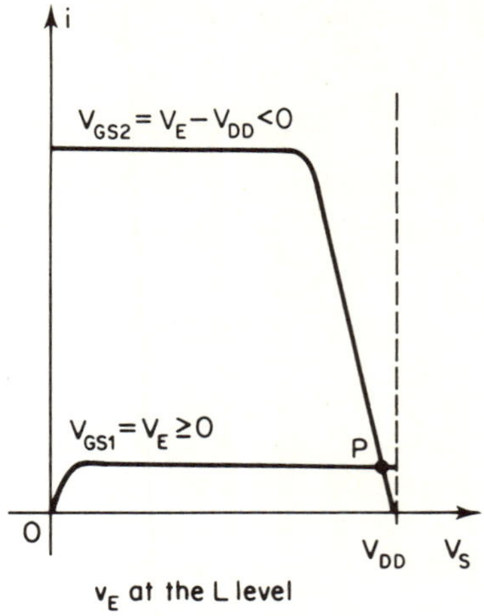

Figure 2.8

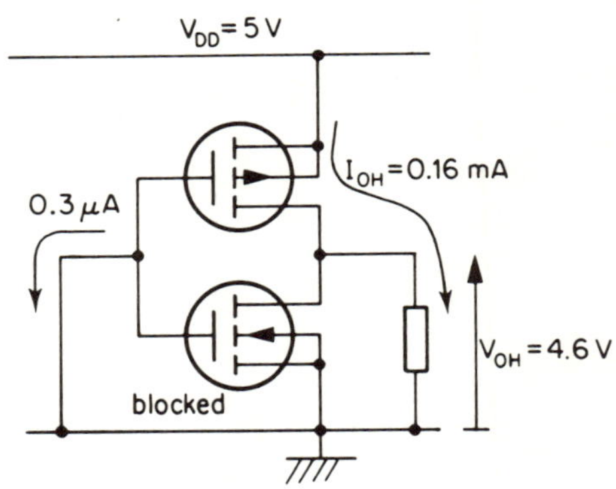

Figure 2.9

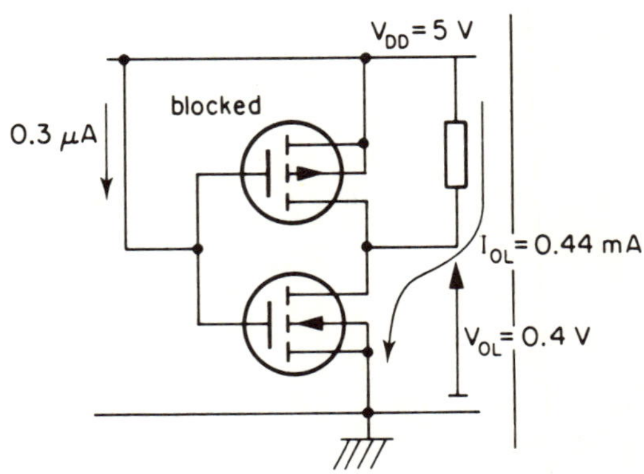

Figure 2.10

M and P are the two operating points characterised by a very small current I.

$$M \text{ corresponds to } V_S \simeq 0$$
$$P \text{ corresponds to } V_S \simeq V_{DD}.$$

Two cases, summarized by Figures 2.9 and 2.10, can occur when $V_{DD} = 5\,V$. These correspond to the B and UB series.

2.2 STATEMENT OF EXERCISES AND PROBLEMS

The exercises and problems which follow are to clarify the concept of *open collector*, *3-state output*, and their applications. These concepts arise from the need to make *connections to a bus.*

Recall that a *bus* is a set of *n* conductors carrying *n* signals which have a global significance. A unidirectional bus is represented in Figure 2.11.

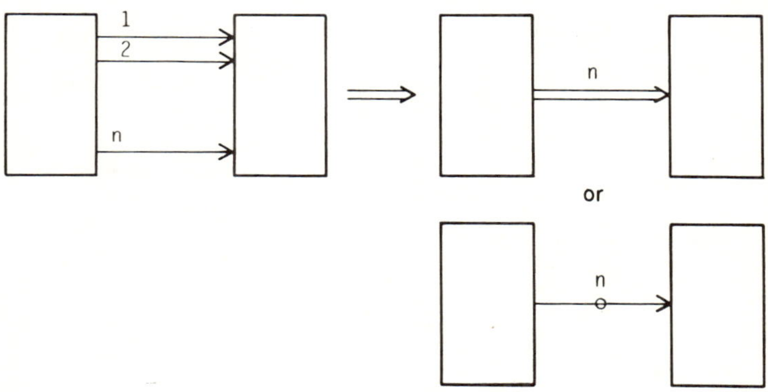

Figure 2.11

A bidirectional bus is represented in Figure 2.12. The arrows indicate the direction of information transfer.

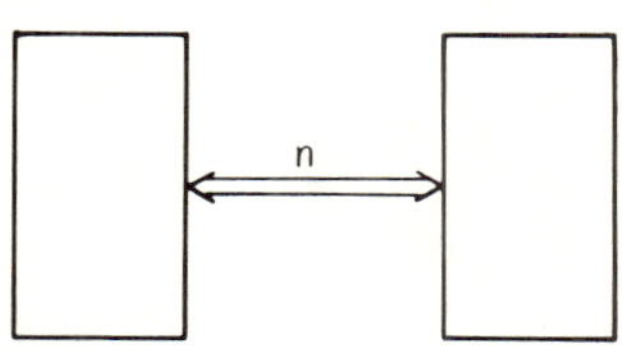

Figure 2.12

Furthermore, a set of signals having different directions of information transfer is also called a bus. Its representation is given in Figure 2.13.

A bus should allow several blocks to exchange information which can be processed at intervals. An architecture suitable for such a system is shown in

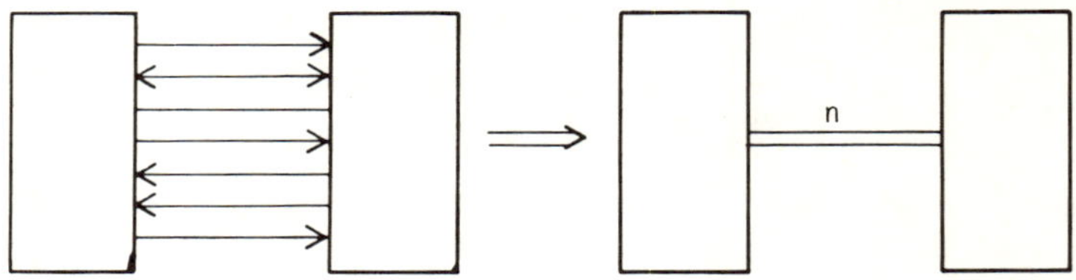

Figure 2.13

Figure 2.14, for the case of two unidirectional buses, and in Figure 2.15 for the case of a bidirectional bus. It is clear (Figure 2.14) that this arrangement involves parallel connection of p outputs.

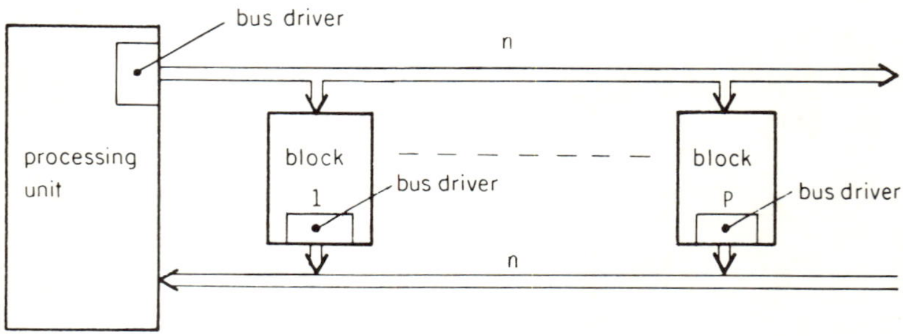

Figure 2.14

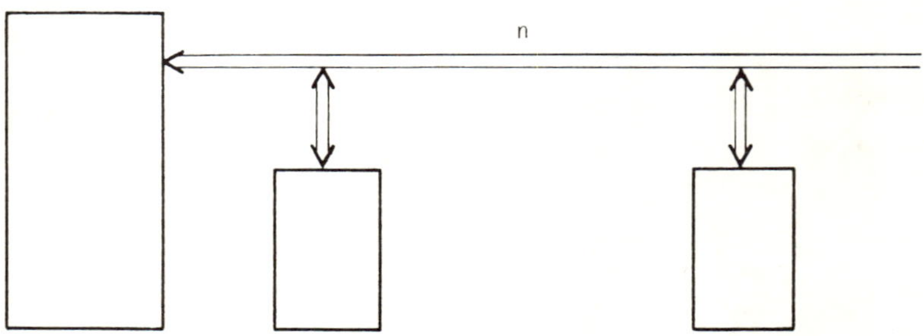

Figure 2.15

Consider one of the n lines and two of the p outputs. The corresponding circuit is given in Figure 2.16.

If the two outputs are not in the same state, current I_{SC} flows, which is not permitted. Therefore the 'totem-pole' configuration does not allow the connection of several outputs.

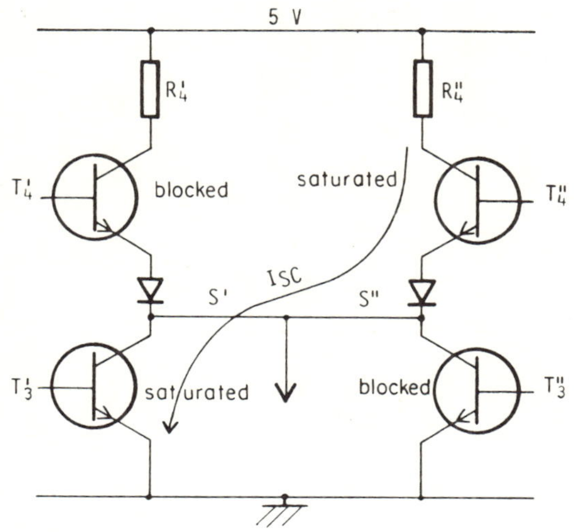

Figure 2.16

2.E.1 Open collector

In order to allow connection of several outputs, the manufacturers produce a circuit, known as an *open collector*, whose circuit is given in Figure 2.17. It can be seen that components R_4, T_4, and D have disappeared.

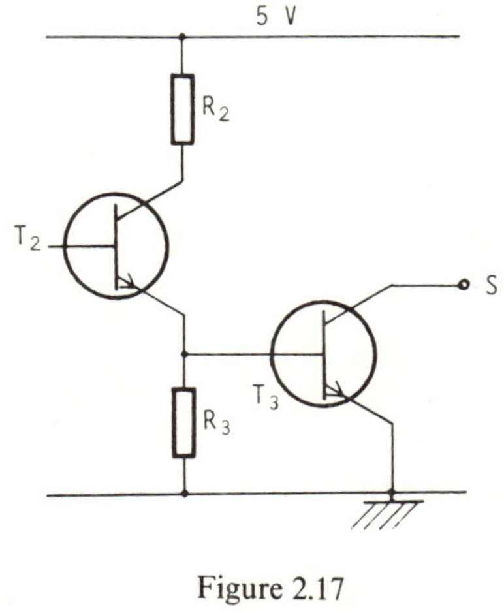

Figure 2.17

Several outputs of this type can be connected together, provided that an external resistor R is connected to 5 V, as shown in Figure 2.18.

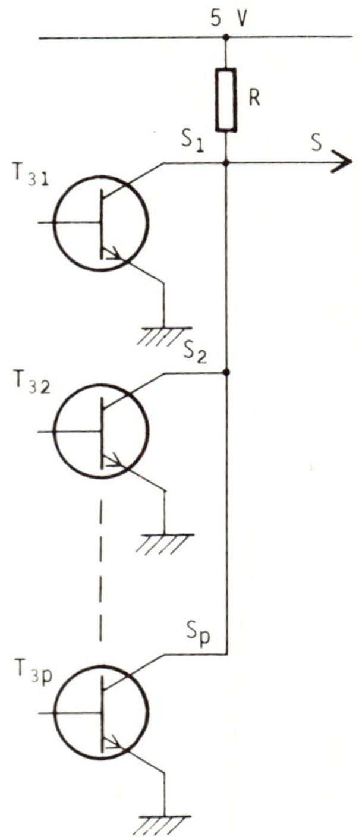

Figure 2.18

Two problems must be resolved:
— the logical equations realized by this connection,
— the value of R.

2.E.1.1 Logical equations

1a. Establish, considering Figure 2.18, in positive logic, the relation:

$$S = f(S_1, S_2, \ldots, S_p).$$

1b. Deduce the logical equations giving S for the different interconnections possible with the 74 series devices indicated in Figures 2.19(a) to 2.19(f).

It should be pointed out that other components can have an open collector output, such as memories, multiplexers, etc.

2.E.1.2 Calculation of R

The manufacturer provides the specifications shown in Figure 2.20.

2a. Let N_S = the number of outputs connected and N_E = the number of inputs connected, each consuming one TTL load; establish the relation between N_S, N_E, and R, ensuring that $V_{OH} = 2.4$ V. Deduce $R = f(N_S, N_E)$.

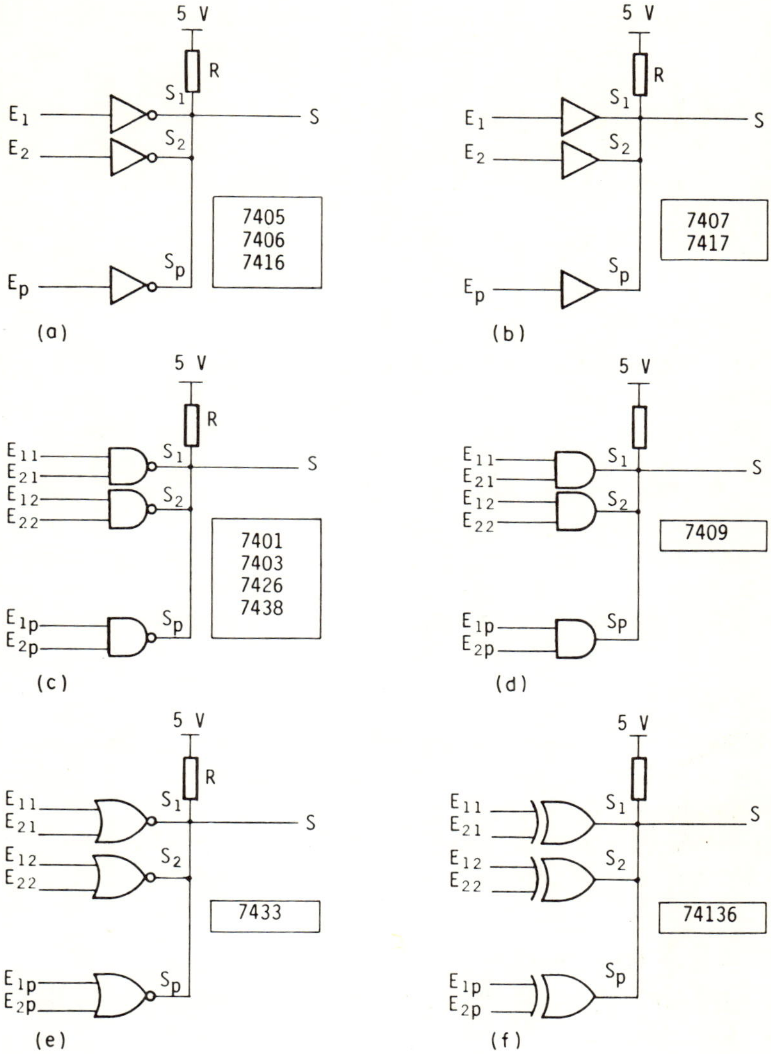

Figure 2.19

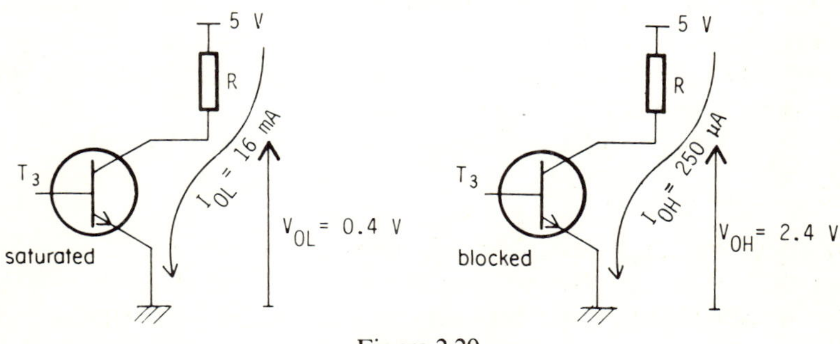

Figure 2.20

2b. With the same notation, establish the relation between N_S,N_E, and R ensuring $V_{OL} = 0.4\ V$. Deduce $R = g(N_S, N_E)$. Neglect the currents of the circuits for which transistor T_3 is blocked.

2c. Write expressions giving R_{max} and R_{min}. Apply these to the case $N_S = 8$, $N_E = 4$.

2d. Represent the two preceding inequalities in the plane $N_E = f(R)$ with N_S as a parameter.

2e. Show that R can also be determined graphically.

2.E.1.3 Applications

3a. Give the circuit of a 2-input multiplexer.

3b. Give the circuit which realizes the structure of Figure 2.14.

3c. Give the circuit which realizes the structure of Figure 2.15.

Comments: This type of connection is called wired-AND, since $S = S_1 \cdot S_2 \cdot \ldots S_p$, sometimes wrongly called 'wired-OR'.

A major disadvantage of this type of connection is that the output impedance in the high state is practically equal to R and can rise to the order of several kilohms; this implies a marked reduction in switching speed.

2.E.2 3-state output

In order to avoid this disadvantage, another structure has been used in order to allow the output stage to operate in any of the three cases represented in Figures 2.2, 2.3, and 2.4.

The circuit of a NAND gate with 3-state output is given in Figure 2.21. The additional elements are indicated.

2.E.2.1 Operation

1a. Show that when the control input $\bar{C}$ is at the low level, the circuit functions normally realizing $S = \overline{E_1 \cdot E_2}$; it is said to be ON.

1b. Show that when the control input $\bar{C}$ is at the high level, the output stage transistors are blocked. The output is therefore at high impedance; the circuit is said to be OFF.

There are 3 possible states for the output, which are summarized in Figures 2.2, 2.3 and 2.4.

1c. Why has transistor T_4 of Figure 2.1 been replaced by transistors T_{41} and T_{42} of Figure 2.21?

2.E.2.2 Interconnection

In the following, we are going to study two circuits, the 74125 and the 74126, which are specially designed to transform 'totem-pole' outputs to 3-state outputs.

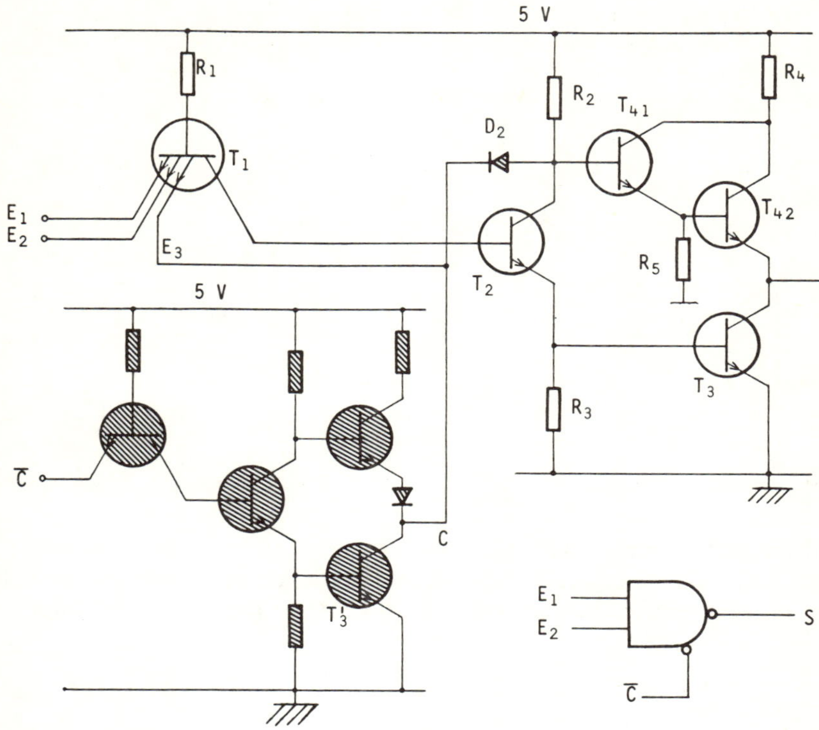

Figure 2.21

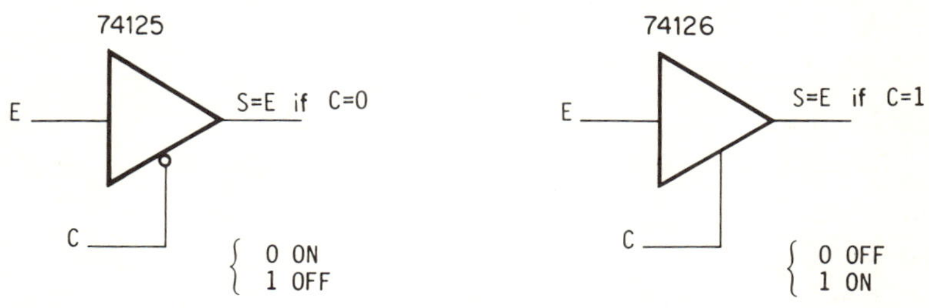

Figure 2.22

They are represented in Figure 2.22 and the manufacturer's specifications are summarized in Figure 2.23.

2a. Let N_S = the number of outputs connected and N_E = the number of inputs fed, each one consuming one TTL load. Establish the relation $N_E = f(N_S)$ when the output is at the high level. Specify, initially, the condition for correct operation which must be verified.

2b. Establish the relation $N_E = g(N_S)$ when the output is at the low level.

2c. Represent the two previous inequalities on the plane of $N_E = f(N_S)$. Show that the problem can be solved graphically.

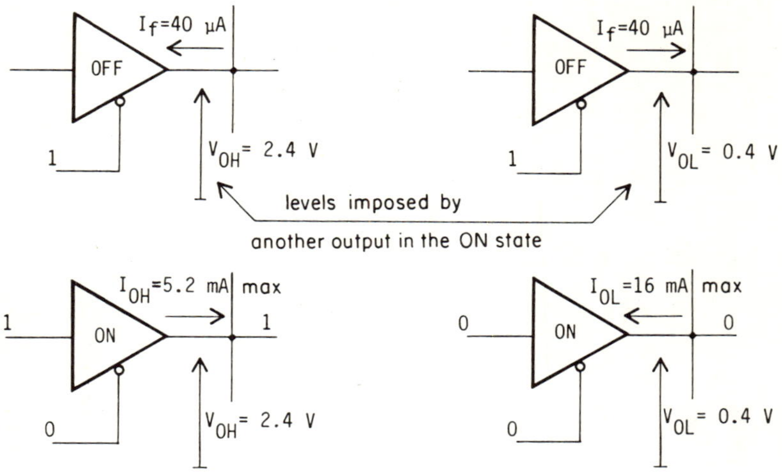

Figure 2.23

2d. The manufacturer defines selection times, t_{ZL} and t_{ZH}, and inhibition times, t_{LZ} and t_{HZ}, according to Figure 2.24. Which conditions must be satisfied this time in order to avoid the situation shown in Figure 2.16 at the time of switching?

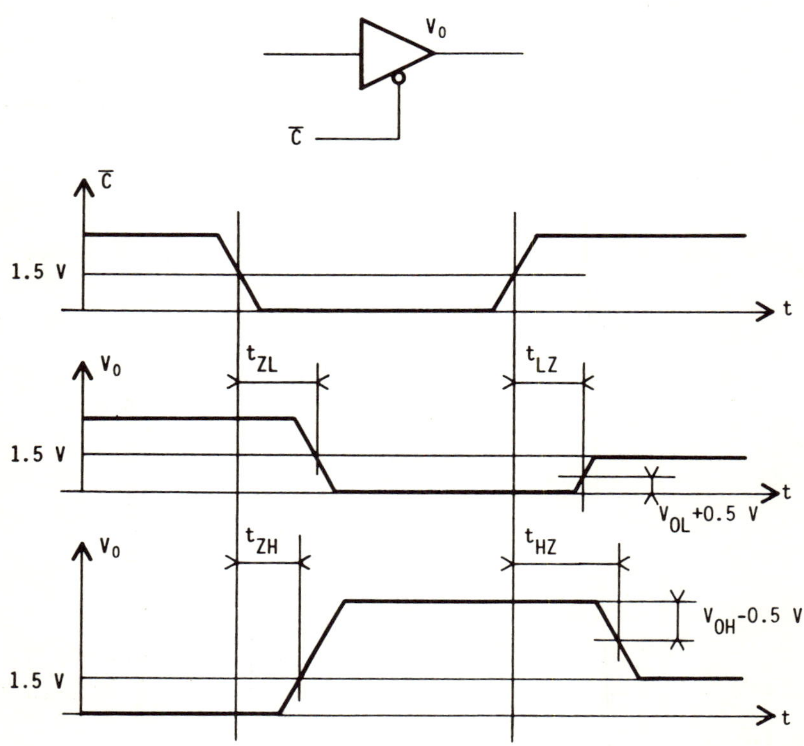

Figure 2.24

2.E.2.3 Applications

3a. Give the circuit of a 2-input multiplexer, realized using the 74125 and 74126 (Figure 2.22). Compare with a classical realization.

3b. Give the circuit which realizes the structure of Figure 2.14.

3c. Give the circuit which realizes the structure of Figure 2.15.

3d. Give the circuit of a bidirectional bus buffer.

2.E.3 Interfacing TTL → CMOS

It is required to control a CMOS circuit, fed from 5 V, by a TTL output as shown in Figure 2.25.

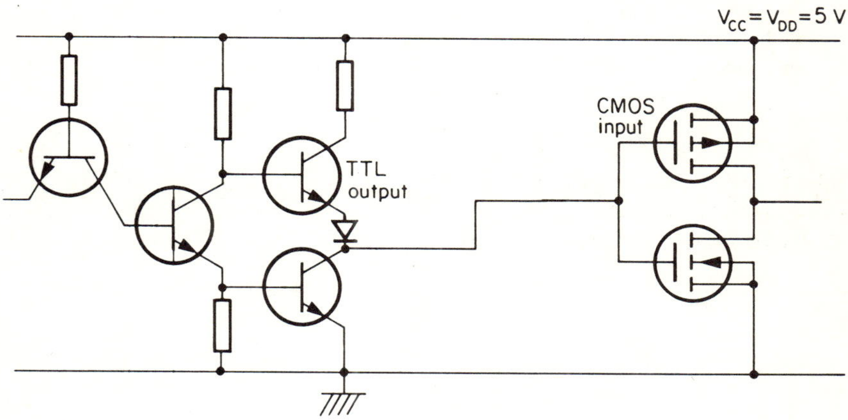

Figure 2.25

Recall the TTL and CMOS family characteristics shown in Figures 2.26 and 2.27.

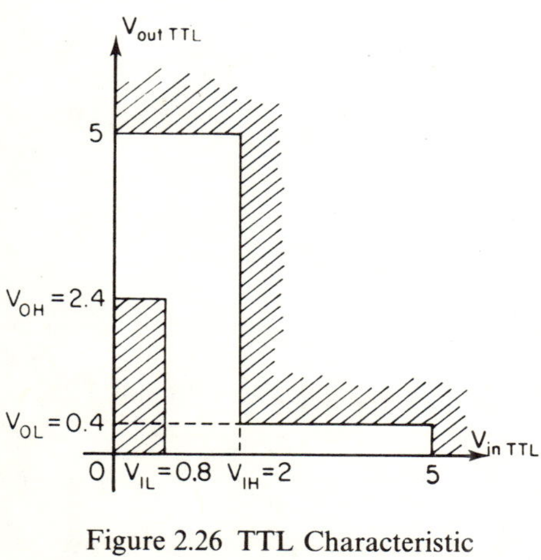

Figure 2.26 TTL Characteristic

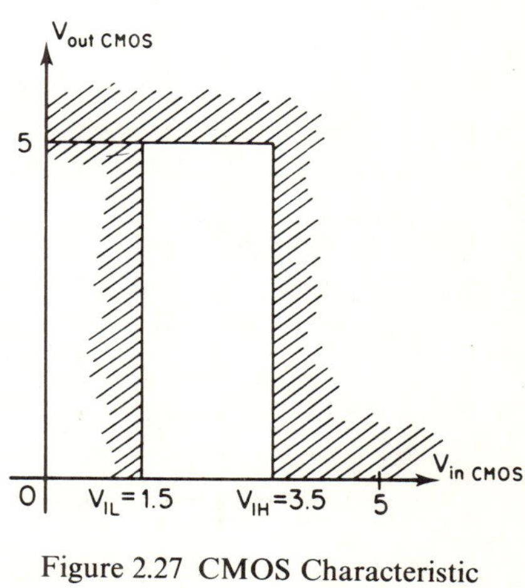

Figure 2.27 CMOS Characteristic

2.E.3.1 TTL output at the L level

Show that there is no problem from either a voltage or current point of view.

2.E.3.2 TTL output at the H level

1. Show that in this case, if $V_{OTTL} < 3.5$ V, operation is incorrect.
2. In order to overcome this problem, a resistance R called a 'pull-up resistor' is connected between the TTL output and $V_{CC} = V_{DD} = 5$ V. Calculate the corresponding value of R in the form of an inequality $R \leqslant R_{max}$ given that in this case $I_{OH} \simeq 100\,\mu$A.
3. Consideration of the permissible current at the L level also requires that $R_{min} \leqslant R$. Calculate R_{min}.
4. Select a value for R.
5. Treat the case of an open collector output.

2.E.4 Interfacing CMOS → TTL

It is required to control a TTL circuit by a CMOS output, fed from 5 V, as shown in Figure 2.28.

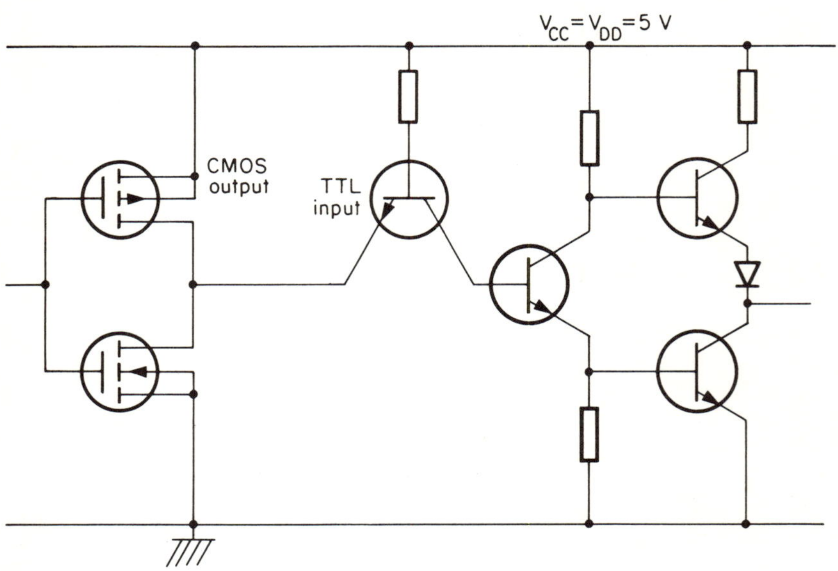

Figure 2.28

The characteristics of the two circuit have been given in Figures 2.26 and 2.27.

2.E.4.1 Voltages

Show that there is no problem when the CMOS output is at the H or L level.

2.E.4.2 Currents

1. Is the CMOS output rated to source or sink the current required by the TTL input? Justify your answer.
2. What can be done about this inadequacy?

2.E.5 CMOS analog switches

This is a basic circuit, like the inverter, and all the other circuits in the family are based on them. The function realized is given in Figure 2.29.

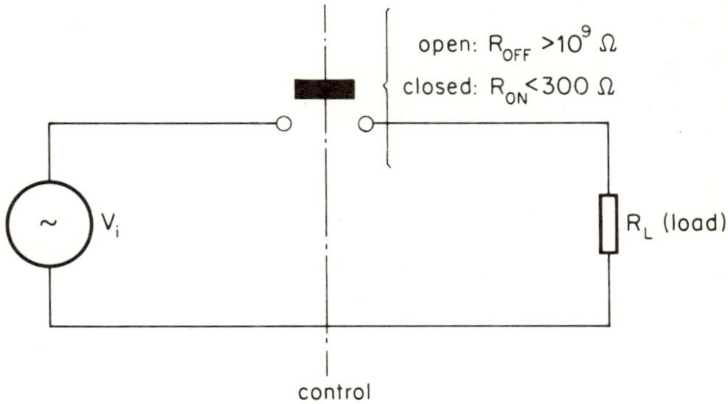

Figure 2.29

Suppose that $-5\,V \leqslant v_i \leqslant +5\,V$

2.E.5.1 N channel switch

The circuit is given in Figure 2.30.

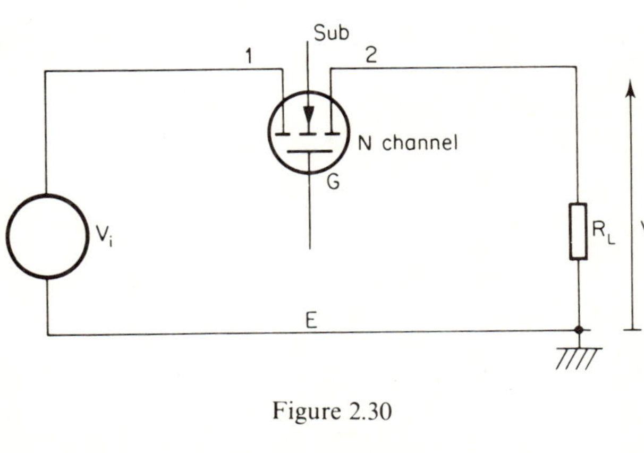

Figure 2.30

1. How is the substrate, Sub, polarized so that the device is not destroyed?
2. What is the polarity of the gate G so that the device is blocked?

3. What is the polarity of the gate G so that the device conducts for $-5\,V \leqslant v_i \leqslant +5\,V$?
This value of v_g will be used subsequently.

4. For $-5\,V \leqslant v_i \leqslant 0$, what is the mode of operation of the transistor, common source or source follower? Replace 1 and 2 by source S and drain D.

5. For $0 \leqslant v_i \leqslant +5\,V$, show that, above a certain value which can be specified, the mode of operation changes if V_{GE} also changes.

6. Sketch the curves $v_0 = f(v_i)$ and $R_{ON} = g(v_i)$.

7. What do you conclude?

2.E.5.2 P channel switch

The circuit is given in Figure 2.31.

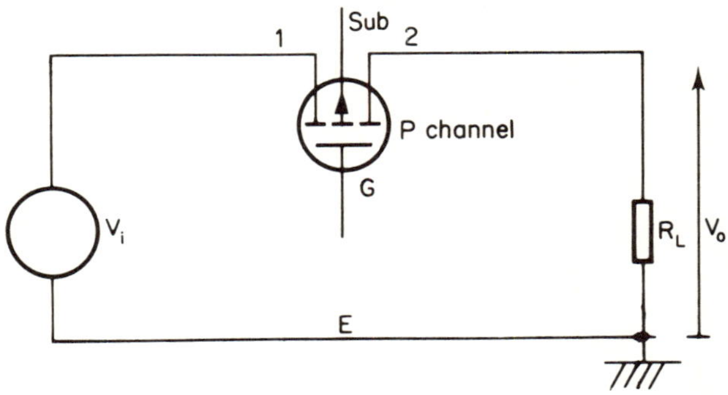

Figure 2.31

Repeat questions 1 to 7 above, adapting the range of variation of v_i.

2.E.5.3. The complete switch

1. Suggest a circuit which overcomes the limitations of the two previous configurations.
2. Analyse the operation of this circuit.
3. Hence deduce the graphs $v_0 = f(v_i)$ and $R_{ON} = g(v_i)$

2.E.5.4 Applications

1. Representing analog switches by the symbol of Figure 2.32, draw diagrams of the principal switching circuits used in practice.

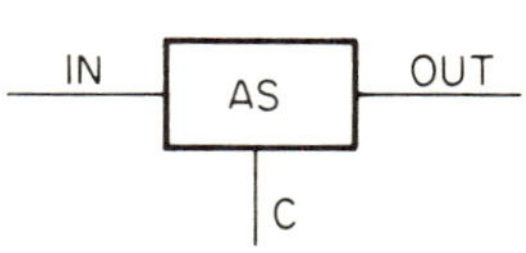

Figure 2.32

2. Draw two diagrams of 3-state outputs of which one uses an analog switch.

3. Explain, with the aid of a timing diagram, the operation of the MC14013B D flip-flop whose circuit is given in Figure 2.33.

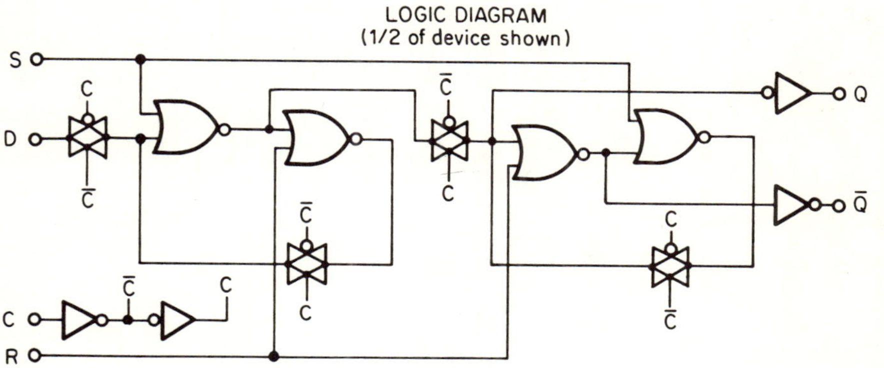

Figure 2.33

2.3 SOLUTIONS

2.S.1 Open collector

2.S.1.1 Logical equations

1a. Transistors T_{31} to T_{3p} can be in different states according to the level of the inputs to the circuit; however, it is clear that the state of the output *S* will be *H* only if all the transistors are blocked.

The table of Figure 2.34 can be set up from which we deduce:

$$S = S_1 \cdot S_2 \cdot \ldots \cdot S_p.$$

S_p	… … … …	S_2	S_1	S
L	… … … …	*L*	*L*	*L*
L	… … … …	*L*	*H*	*L*
L	… … … …	*H*	*L*	*L*
⋮		⋮	⋮	⋮
H	… … … …	*H*	*L*	*L*
H	… … … …	*H*	*H*	*H*

Figure 2.34

1b. Figure 2.19(a) allows us to write:

$$S_1 = \overline{E_1}, \quad S_2 = \overline{E_2}, \ldots, \quad S_p = \overline{E_p}, \quad \text{from which}$$
$$S \;= \overline{E_1} \cdot \overline{E_2} \cdot \cdots \overline{E_p} = \overline{E_1 + E_2 + \cdots + E_p}.$$

Similarly, from Figure 2.19(b):

$$S_1 = E_1, \quad S_2 = E_2, \ldots, \quad S_p = E_p, \quad \text{from which}$$
$$S \;= E_1 \cdot E_2 \cdot \cdots \cdot E_p;$$

from Figure 2.19(c):

$$S_1 = \overline{E_{11} \cdot E_{21}}, \quad S_2 = \overline{E_{12} \cdot E_{22}}, \ldots, \quad S_p = \overline{E_{1p} \cdot E_{2p}}, \quad \text{from which}$$
$$S = \overline{E_{11} \cdot E_{21}} \cdot \overline{E_{12} \cdot E_{22}} \cdot \ldots \cdot \overline{E_{1p} \cdot E_{2p}}$$
$$= \overline{E_{11} \cdot E_{21} + E_{12} \cdot E_{22} + \cdots + E_{1p} \cdot E_{2p}};$$

from Figure 2.19(d):

$$S_1 = E_{11} \cdot E_{21}, \quad S_2 = E_{12} \cdot E_{22}, \ldots, \quad S_p = E_{1p} \cdot E_{2p}, \quad \text{from which}$$
$$S \;= E_{11} \cdot E_{21} \cdot E_{12} \cdot E_{22} \cdot \ldots \cdot E_{1p} \cdot E_{2p};$$

from Figure 2.19(e):

$$S_1 = \overline{E_{11} + E_{21}}, \quad S_2 = \overline{E_{12} + E_{22}}, \ldots, \quad S_p = \overline{E_{1p} + E_{2p}}, \quad \text{from which}$$
$$S = \overline{E_{11} + E_{21}} \cdot \overline{E_{12} + E_{22}} \cdot \ldots \cdot \overline{E_{1p} + E_{2p}}$$
$$= \overline{E_{11} + E_{21} + E_{12} E_{22} + \cdots + E_{1p} + E_{2p}};$$

from Figure 2.19(f):

$$S_1 = E_{11} \oplus E_{21}, \quad S_2 = E_{12} \oplus E_{22}, \ldots, \quad S_p = E_{1p} \oplus E_{2p}$$
$$S \;= (E_{11} \oplus E_{21}) \cdot (E_{12} \oplus E_{22}) \cdot \ldots \cdot (E_{1p} \oplus E_{2p}).$$

2.S.1.2. Calculation of R

2a. The corresponding diagram is given in Figure 2.35. We can write (with R in $k\Omega$):

$$5 - R(0.25\,N_S + 0.04\,N_E) \geqslant 2.4 \tag{1}$$

from which
$$R \leqslant \frac{5 - 2.4}{0.25 N_S + 0.04 N_E} \tag{2}$$

2b. The corresponding diagram is given in Figure 2.36. We can write (with R in $k\Omega$):

$$\frac{5 - 0.04}{\mathrm{R}} + 1.6\,N_E \leqslant 16 \tag{3}$$

from which
$$R \geqslant \frac{5 - 0.4}{16 - 1.6\,N_E} \tag{4}$$

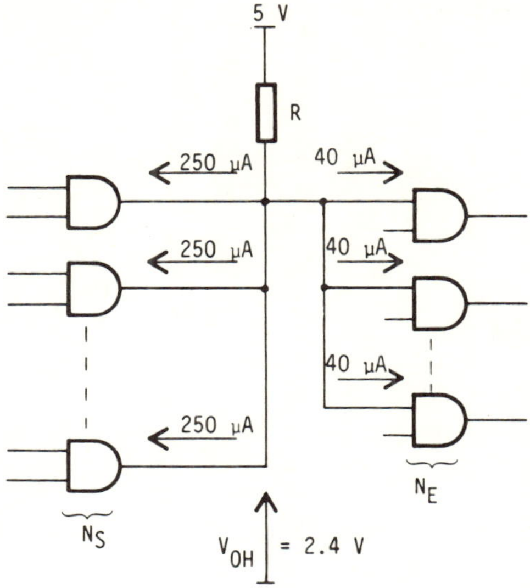

Figure 2.35

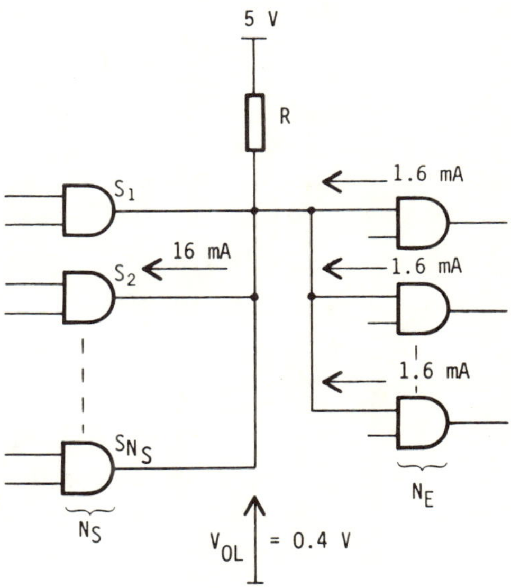

Figure 2.36

2c. From expression (2):

$$R_{\max} = \frac{2.6}{0.25 N_S + 0.04 N_E}$$

from expression (4):

$$R_{\text{min}} = \frac{4.6}{16 - 1.6N_{\text{E}}}$$

In the application, $N_S = 8$, $N_E = 4$, from which

$$R_{\text{max}} = \frac{2.6}{8 \times 0.25 + 4 \times 0.04} \simeq 1.2\text{k}\Omega$$

$$R_{\text{min}} = \frac{4.6}{16 - 4 \times 1.6} \simeq 0.479\, k\Omega$$

hence $479\,\Omega \leqslant R \leqslant 1200\,\Omega$

The choice of a preferred value between these limits depends on the requirements of power dissipation and switching speed.

Comment: If the calculation gives $R_{\text{min}} > R_{\text{max}}$ or $R_{\text{min}} < 0$, this shows that the proposed connection is not realizable.

2d. From expressions (1) and (3):

$$N_E \leqslant 65/R - 6.25\, N_S,$$

which corresponds to a family of curves with N_S as a parameter,

and:

$$N_E \leqslant 10 - 2.875/R,$$

which corresponds to a curve which bounds the previous family.

The corresponding graphs are given in Figure 2.37.

2e. The graphical approach is simple. Reconsider the previous example. For $N_E = 4$, the corresponding horizontal line is selected (Figure 2.38):

its intersection with curve (1) gives $R_{\text{min}} \simeq 500\,\Omega$

its intersection with curve (3) gives $R_{\text{max}} \simeq 1\,150\,\Omega$

The possible values of R are represented by the segment AB.

2.S.1.3 Applications

3a. We know that $S = \bar{A}\cdot E_0 + A\cdot E_1$ (see Section 1.E.3). The structure of Figure 2.19(c), realized with two outputs connected, gives:

$$S = \overline{E_{11}\cdot E_{21} + E_{12}\cdot E_{22}}$$

Therefore it is sufficient to make $E_{11} = E_0$, $E_{21} = \bar{A}$, $E_{12} = E_1$, and $E_{22} = A$, from which the diagram of Figure 2.39 can be drawn.

3b. The diagram is given in Figure 2.40. The n resistances R can be calculated as shown above.

3c. The diagram is given in Figure 2.41. A single group of the n validation inputs is in the '1' state, the others are all in the '0' state in order to block the output transistors T_3.

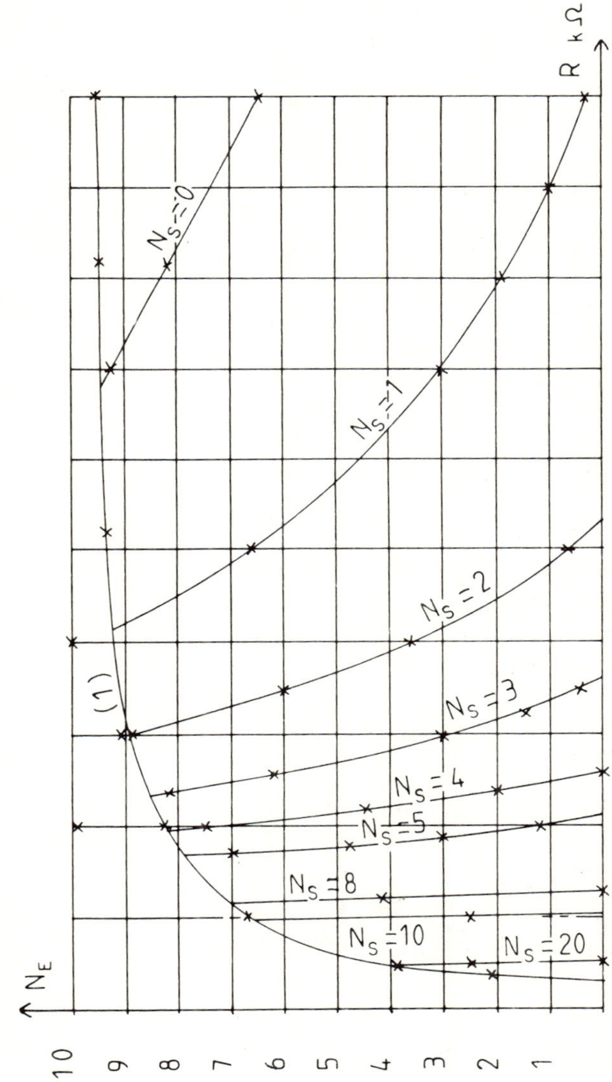
N_E
R kΩ
10
9
8
7
6
5
4
3
2
1
1
5
10
N_S = 0
N_S = 1
N_S = 2
N_S = 3
N_S = 4
N_S = 5
N_S = 8
N_S = 10
N_S = 20
(1)

Figure 2.37

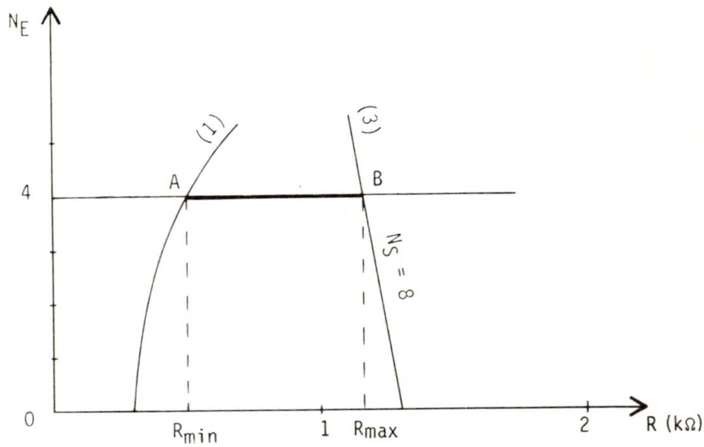

Figure 2.38

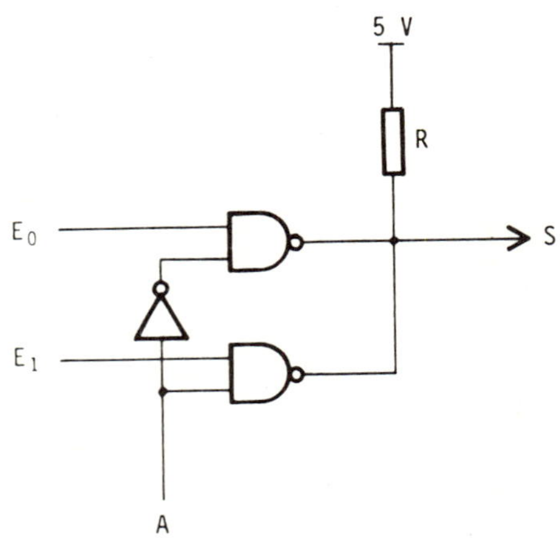

Figure 2.39

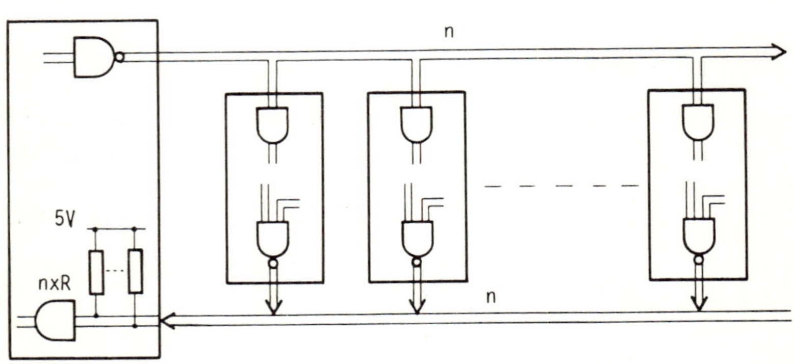

Figure 2.40

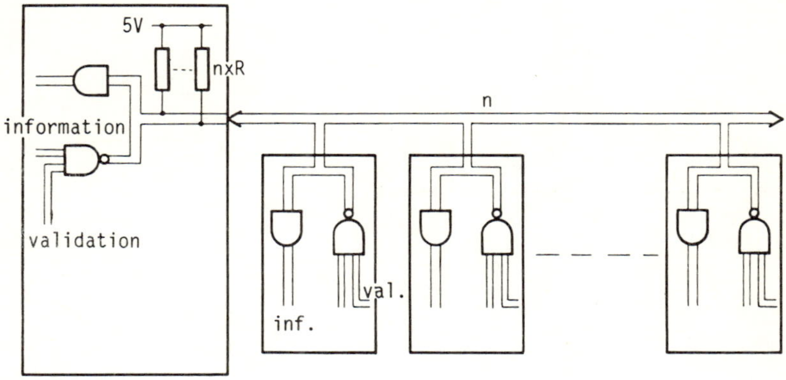

Figure 2.41

2.S.2 3-state output

2.S.2.1 Operation

1a. When $\overline{C}$ is at a low level, the output C of the inverter is at a high level:

— the emitter E_3 is disabled;
— diode D_2 is reverse biassed.

Consequently the circuit operates normally: $S = \overline{E_1 \cdot E_2}$.

1b. When $\overline{C}$ is at a high level, the output C of the inverter is at a low level:
— The current in transistor T_1 flows to earth via E_3 and T'_3, consequently E_1 and E_2 have no effect;
— T_2 and T_3 are blocked;
— diode D_2 conducts which turns T_{41} and T_{42} off.
The output stage is therefore in the state indicated in Figure 2.4.

1c. T_{41} and T_{42} comprise a Darlington type current amplifier. It provides a substantial current at the high level, as will be seen.

2.S.2.2 Interconnection

2a. In order to avoid the situation shown in Figure 2.16, only one of the N_S circuits is ON. At the high level it can provide 5.2 mA. Figure 2.42 shows this interconnection.
The activated output can drive 5 200/40 = 130 TTL loads. It is clear that:

$$[(N_S - 1) + N_E]0.04 \leqslant 5.2$$

$$\text{or} \quad N_E \leqslant 130 - (N_S - 1), \tag{1}$$

$$\text{and} \quad N_E \leqslant 131 - N_S.$$

2b. Figure 2.43 shows the situation.

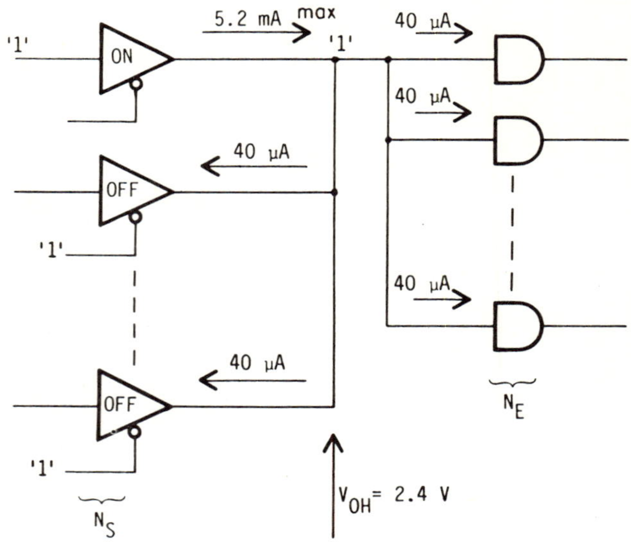

Figure 2.42

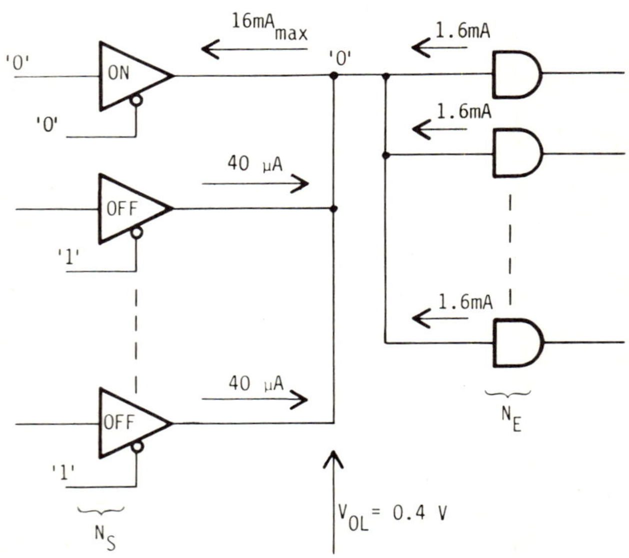

Figure 2.43

The following inequality should be verified:

$$0.04(N_S - 1) + 1.6N_E \leqslant 16,$$

$$\text{or} \quad N_E \leqslant 10 - 0.025(N_S - 1). \tag{2}$$

2c. The graphical representation is given in Figure 2.44. For a given N_E,

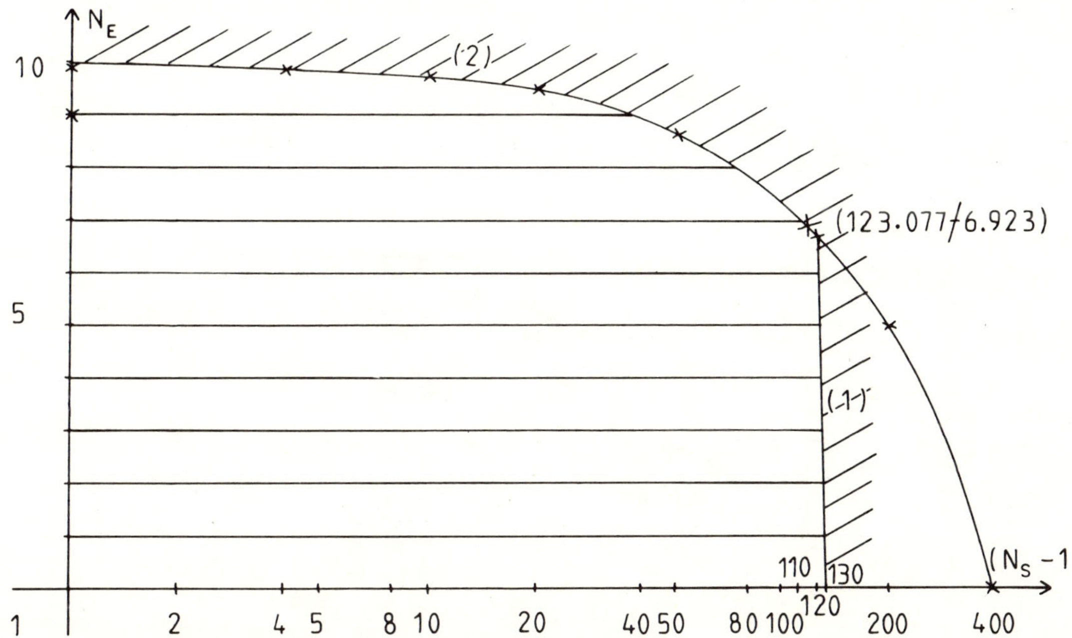

N_E
10
5
1
2
4
5
8
10
20
40
50
80
100
110
120
130
200
400
$(N_S - 1)$
(2)
(1)
(123.077/6.923)

Figure 2.44

N_{Smax} is given by the intersection of the corresponding horizontal and curve (2) or (1).

2d. In order to avoid the situation shown in Figure 2.16 during switching, the manufacturer ensures that the inhibition times (put into the OFF state) are shorter than the selection times (put into the ON state). Therefore:

$$t_{LZ} \text{ and } t_{HZ} < t_{ZL} \text{ and } t_{ZH}.$$

2.S.2.3 Applications

3a. The very simple diagram is given in Figure 2.45.

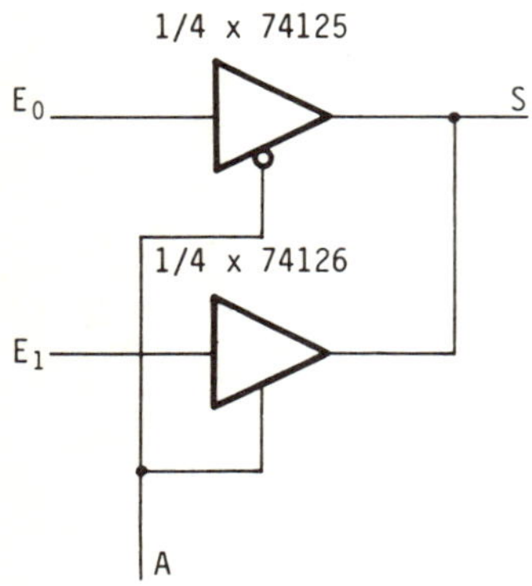

Figure 2.45

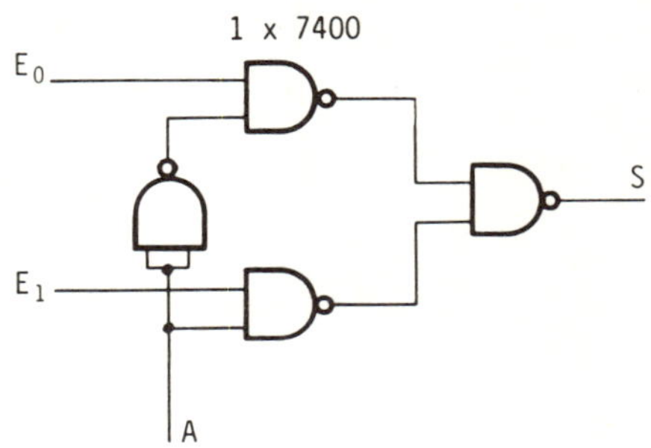

Figure 2.46

Recall the classical solution (Figure 2.46). It requires more hardware, for example a 4 × 2-input multiplexer requires one 74125 package and one 74126 package or four 7400 packages.

3b. The diagram is given in Figure 2.47. Notice that there are no resistances; only one of the 3-state output circuits must be ON.

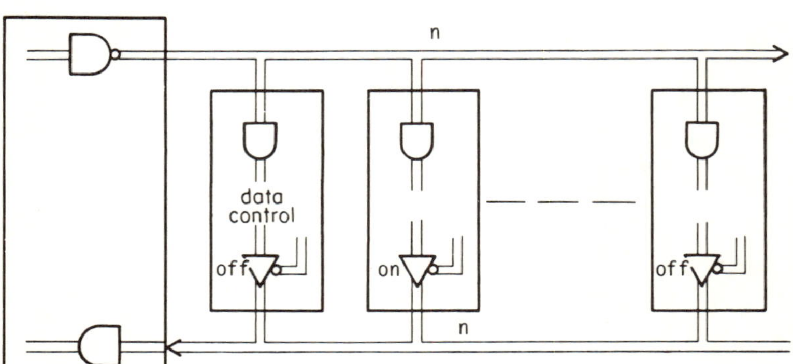

Figure 2.47

3c. The diagram is given in Figure 2.48. Again resistances are not used; as before, only one of the 3-state output circuits must be ON.

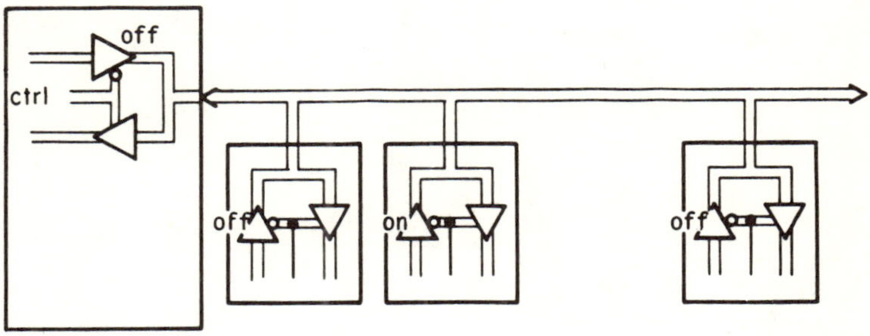

Figure 2.48

3d. The diagram is given in Figures 2.49 and 2.50 for the two directions of information transfer.

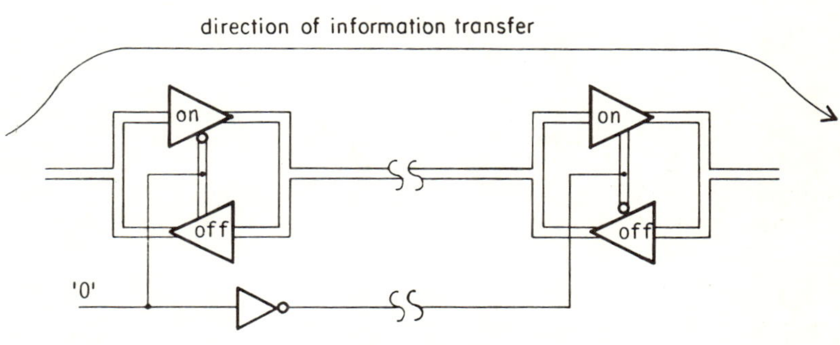

Figure 2.49

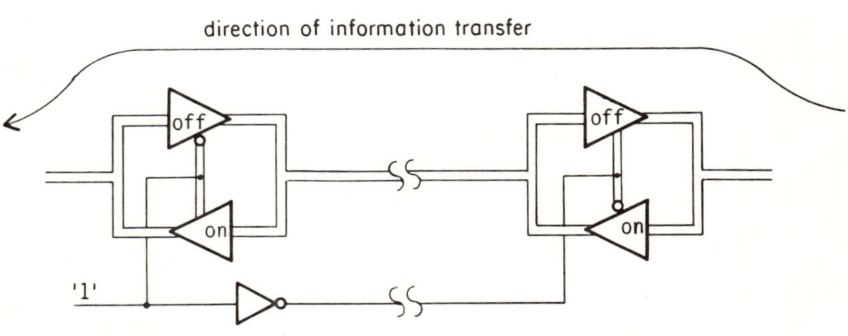

Figure 2.50

2.S.3 Interfacing TTL→CMOS

2.S.3.1 TTL output at the L level

Noticing that $V_{OTTL} = V_{1CMOS}$, the two characteristics can be combined as shown in Figure 2.51.

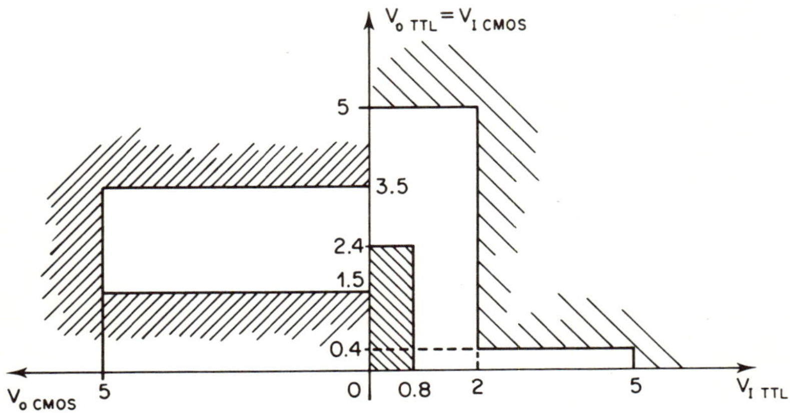

Figure 2.51

It is clear that if $2 \leqslant v_{1TTL} \leqslant 5$ then $v_{OTTL} \leqslant 0.4$ and $v_{OCMOS} = 5\,V$.

Therefore an *H* level at the TTL input corresponds to an *H* level at the CMOS output, the CMOS input current being negligible.

2.S.3.2 TTL output at the H level

1. Figure 2.51 shows that if $0 \leqslant v_{1TTL} \leqslant 0.8\,V$, the TTL output will be such that $2.4\,V \leqslant v_{OTTL} = v_{1CMOS} \leqslant 5\,V$.

If $3.5\,V \leqslant v_{OTTL} = v_{1CMOS} \leqslant 5\,V$, then $v_{OCMOS} = 0\,V$.

Therefore an *L* level at the TTL input corresponds to an *L* level at the CMOS output.

If $2.4\,V \leqslant v_{OTTL} = v_{1CMOS} \leqslant 3.5\,V$, the state of the CMOS output may be *L* or *H* according to the actual position on the transfer characteristic, as shown in Figure 2.52.

2. The arrangement is given in Figure 2.53.

The inequality $5 - RI_{OH} \geqslant 3.5$ should be verified.

It can be deduced that $R \leqslant \dfrac{5 - 3.5}{100 \times 10^{-6}} = 15\,000\,\Omega$

from which $R_{max} = 15\,k\Omega$.

3. It is also necessary to check that at the *L* level, the TTL output current does not exceed $I_{OL} = 16\,mA$ as shown in Figure 2.54.

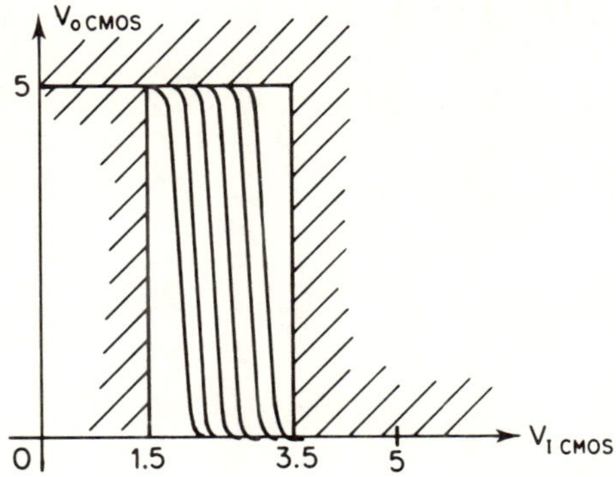

Figure 2.52

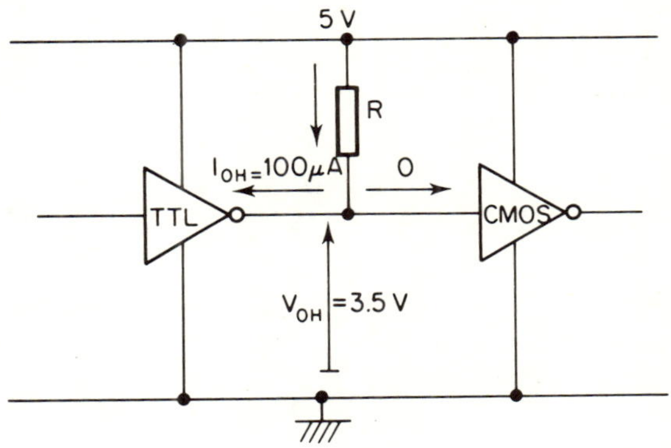

Figure 2.53

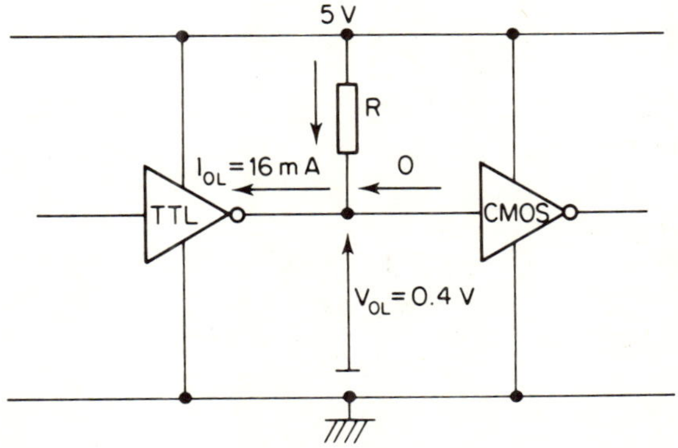

Figure 2.54

The inequality $5 - RI_{OL} \leqslant 0.4$ should be verified.

It can be deduced that $\dfrac{5 - 0.4}{16 \times 10^{-3}} = 330\,\Omega \leqslant R$, from which $R_{min} = 330\,\Omega$.

4. These results are summarized in Figure 2.55.

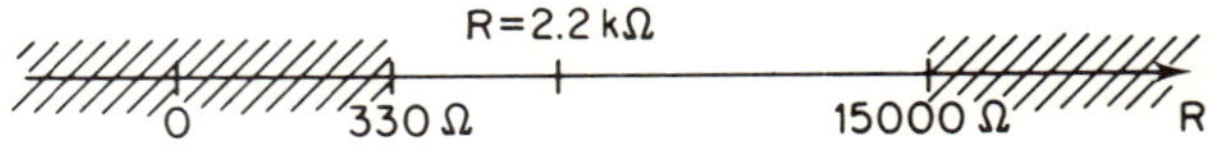

Figure 2.55

We shall choose the standard value of 2.2 kΩ which offers a good compromise between speed and power consumption.

5. If the TTL output has an open collector, I_{OH} is specified as 250 μA which gives $R_{max} = 6\,000\,\Omega$.

This does not change the choice of 4 in any way.

2.S.4. Interfacing CMOS → TTL

2.S.4.1 Voltages

Noticing that $V_{0CMOS} = V_{1TTL}$, the two characteristics can be combined as shown in Figure 2.56.

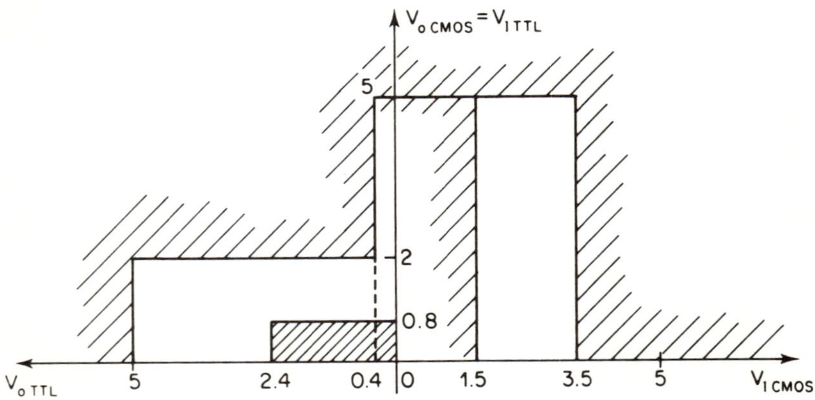

Figure 2.56

It is clear that the CMOS output can take only the values 0 and 5 V, which correspond perfectly with the *L* and *H* levels of TTL technology.

2.S.4.2 Currents

1. Figures 2.9 and 2.10 show that a CMOS output is unable to accept $I_{IL} = 1.6\,\text{mA}$ from a TTL input when $V_{OL} = 0.4\,\text{V}$.

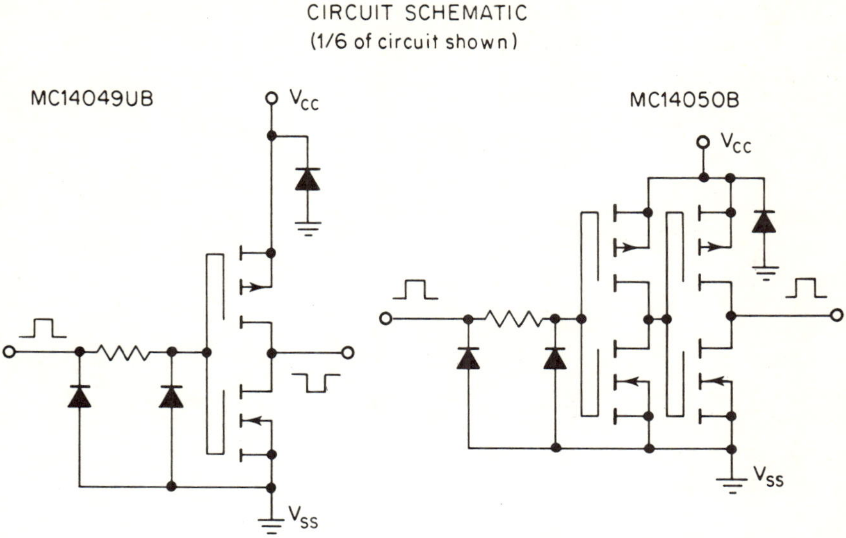

Figure 2.57

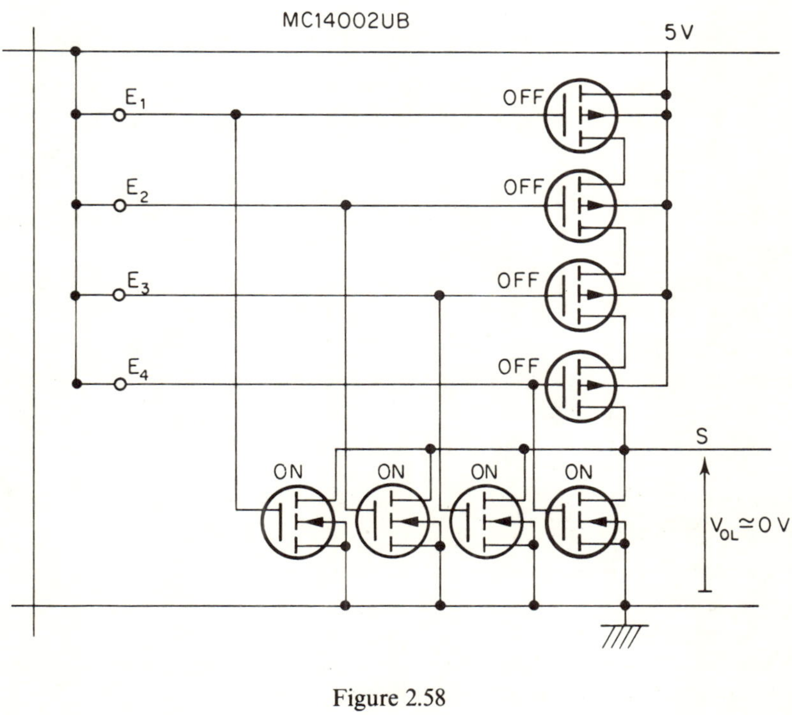

Figure 2.58

2. This inadequacy is avoided by the use of CMOS amplifier circuits type MC14049UB and MC14050B (Figure 2.57) which accept:

$$I_{OL} = 3.2\,\text{mA at } V_{OL} = 0.4\,\text{V}$$
$$\text{and}\quad I_{OH} = -2.5\,\text{mA at } V_{OH} = 2.5\,\text{V}$$

Another solution is to use an MC140020UB NOR gate with the four inputs connected as shown in Figure 2.58. An MC14002B is also suitable.

Each ON transistor absorbs 0.44 mA giving a total of 1.76 mA which corresponds to more than one TTL load of 1.6 mA.

The circuit is equivalent to an inverter, as shown in Figure 2.59.

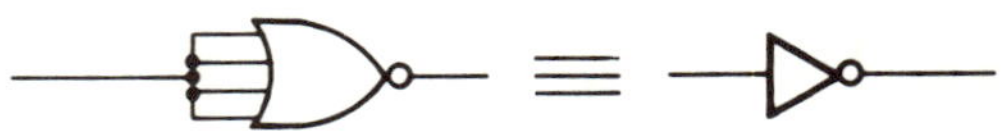

Figure 2.59

2.S.5 CMOS analog switches

2.S.5.1 N channel switch

1. Since the substrate is *P* type, it must be polarized to the lowest voltage of the circuit, so that the substrate–source and substrate–drain diodes never conduct.

$v_{imin} = -5\,\text{V}$ so the substrate will be polarized to this value.

2. It is clear that if $V_{GE} = -5\,\text{V}$, the transistor will always be blocked for all values of V_{in} since G will never be positive with respect to 1 or 2.

3. On the other hand, if $V_{GE} = +5\,\text{V}$, the transistor will be liable to conduct according to the value of v_i.

4. When $-5\,\text{V} \leqslant v_i \leqslant 0$ the transistor operates in common source, since the voltage v_{GS} varies between $+5$ and $+10\,\text{V}$.
The transistor chooses 1 as the source and 2 as the drain. If $R_L \gg R_{ON}$ then $v_0 \simeq v_i$.

5. When $0 \leqslant v_i \leqslant +5\,\text{V}$ the voltage v_{GS} takes the value $v_{GS} = 5 - v_i$. The transistor operates as before until $v_{GS} = V_{TH} \simeq 2\,\text{V}$, called the *threshold voltage*, from which $V_{imax} = 3\,\text{V}$. Above this value, the transistor chooses 1 as the drain, 2 as the source, and operates as a source follower if $V_{GE} = V_{TH} + v_{in}$.

6. The curves $v_0 = f(v_i)$ and $R_{ON} = g(v_i)$ are given in Figure 2.60.

2.S.5.2 P channel switch

1. The previous reasoning leads to a substrate polarisation of $+5\,\text{V}$.
2. The transistor will always be blocked for $V_{GE} = +5\,\text{V}$.
3. The transistor will be liable to conduct if $V_{GE} = -5\,\text{V}$.
4. When $0 \leqslant v_i \leqslant +5\,\text{V}$, the transistor operates, in common source, it chooses 1 as the source and 2 as the drain.

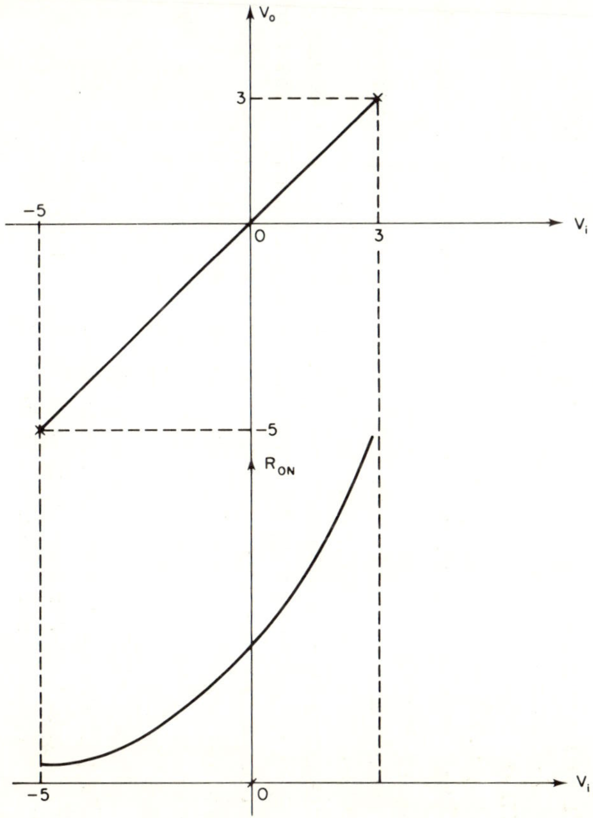

Figure 2.60

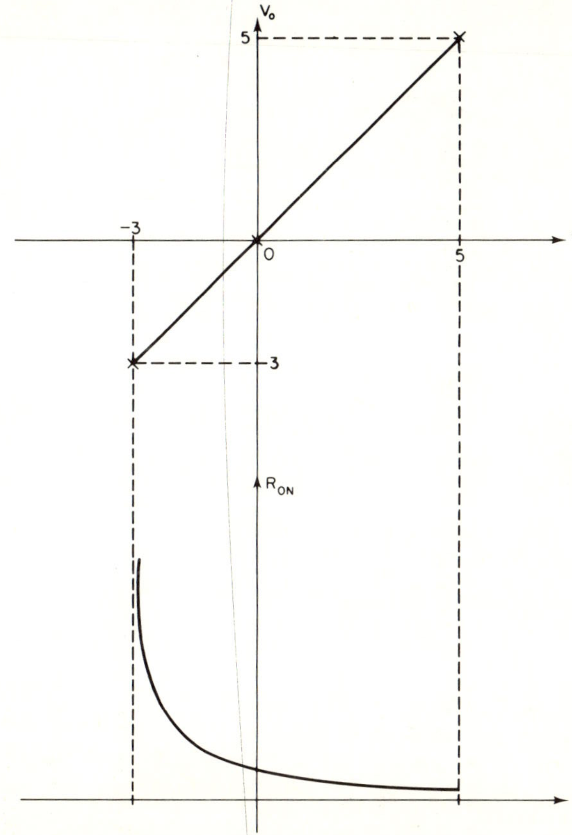

Figure 2.61

5. When $-5\,V \leqslant v_i \leqslant 0$, operation in common source occurs until $v_{GS} = V_{TH} = -2\,V$. This occurs for $V_{imin} = -3\,V$.

6. The curves $v_0 = f(v_i)$ and $R_{ON} = g(v_i)$ are given in Figure 2.61.

2.S.5.3 The complete switch

1. The configuration shown in Figure 2.62 consists of *N* channel and *P* channel transistors connected in parallel.

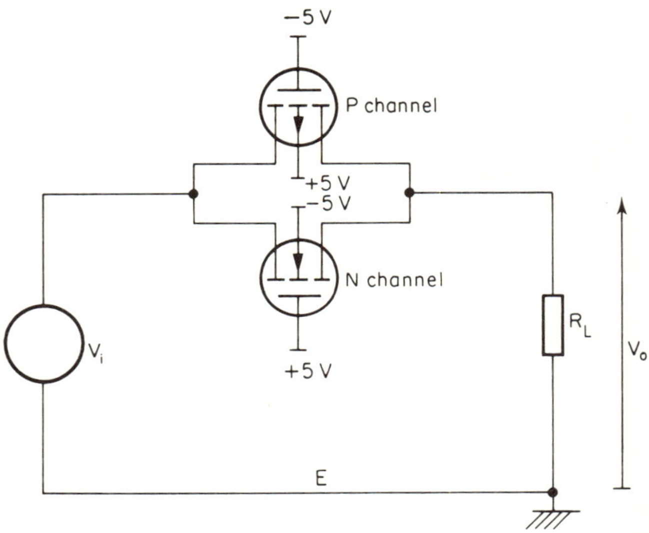

Figure 2.62

2. For $-5\,V \leqslant v_i \leqslant -3\,V$ the *N* channel transistor operates in common source and conceals the operation of the *P* channel transistor.

For $+3\,V \leqslant v_i \leqslant +5\,V$ the roles of the two transistors are exchanged.

For $-3\,V \leqslant v_i \leqslant +3\,V$ the two transistors share the current.

3. The curves given in Figure 2.63 are obtained from Figures 2.60 and 2.61. The asymmetry of the curves $R_{ON} = f(v_i)$ causes a maximum in R_{ON}. In order to avoid this shortcoming, the polarization of the substrate of the *N* channel transistor is controlled from v_i with the help of an additional *P* channel transistor, as shown in Figure 2.64. The R_{ON} curve of Figure 2.63 is modified as shown in Figure 2.65.

The definitive diagram and symbol are given in Figure 2.66.

The circuit can operate in a unipolar mode:

$$\text{with } V_{SS} = 0 \quad \text{and} \quad V_{DD} = 15\,V, \quad \text{therefore} \quad 0 \leqslant V_{in} \leqslant 15\,V$$

or in a bipolar mode:

$$\text{with } V_{SS} = -7.5\,V \quad \text{and} \quad V_{DD} = +7.5\,V, \quad \text{threfore} \quad -7.5\,V \leqslant V_{in} \leqslant +7.5\,V$$

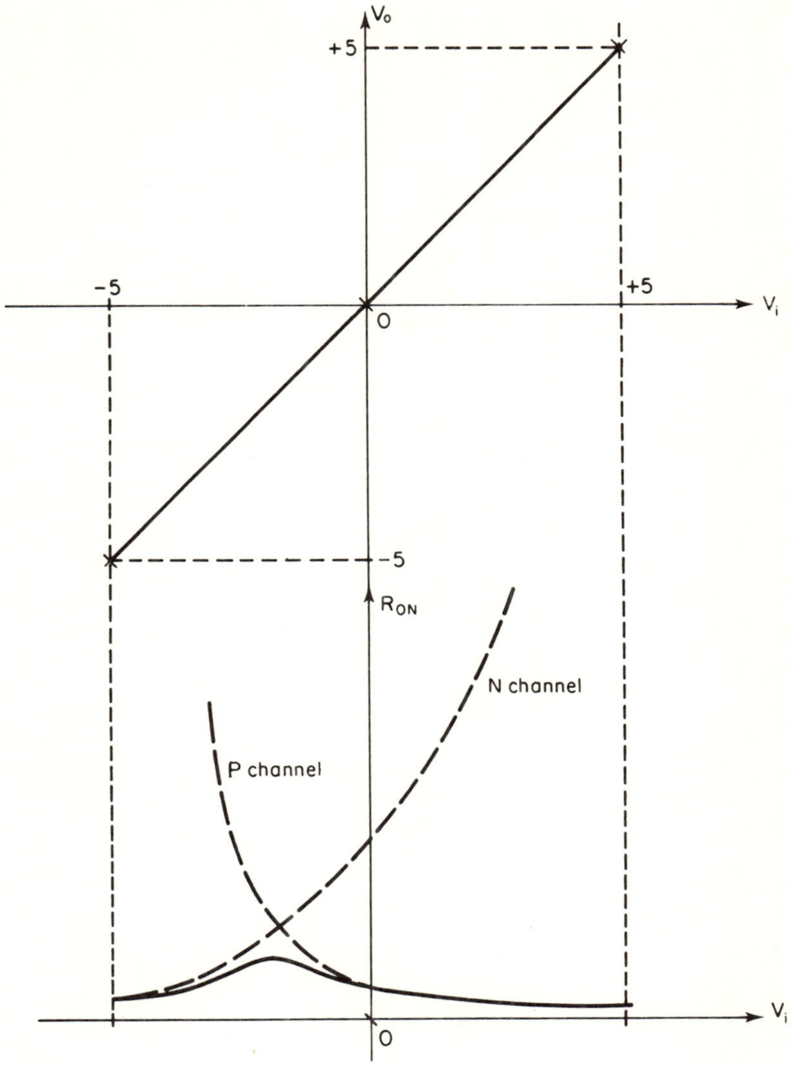

Figure 2.63

2.S.5.4 Applications

1. Figure 2.67 shows the principal types of switching circuits.
2. The two diagrams for 3-state output are given in Figures 2.68 and 2.69.

3a. Figure 2.70 summarizes the operation of the analog switch.

3b. Let us set the clock C to 'O' and force R to '1' in order to find the initial state, as shown in Figure 2.71.

Changing R to '0' does not change the general state of the flip-flop.

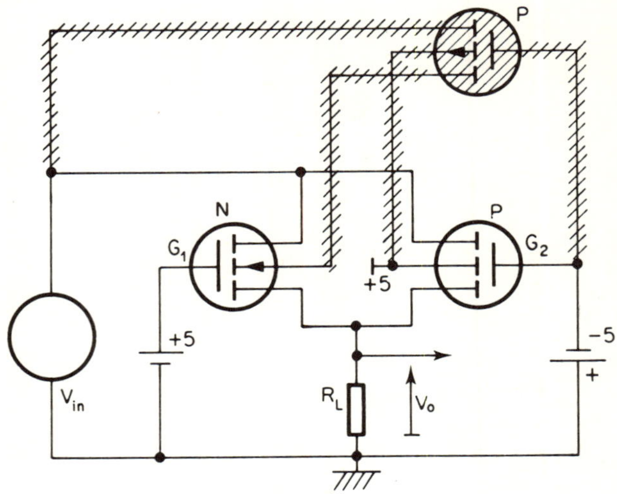

Figure 2.64

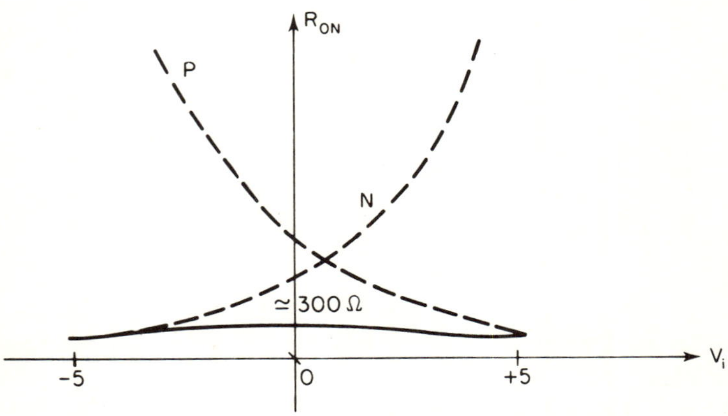

Figure 2.65

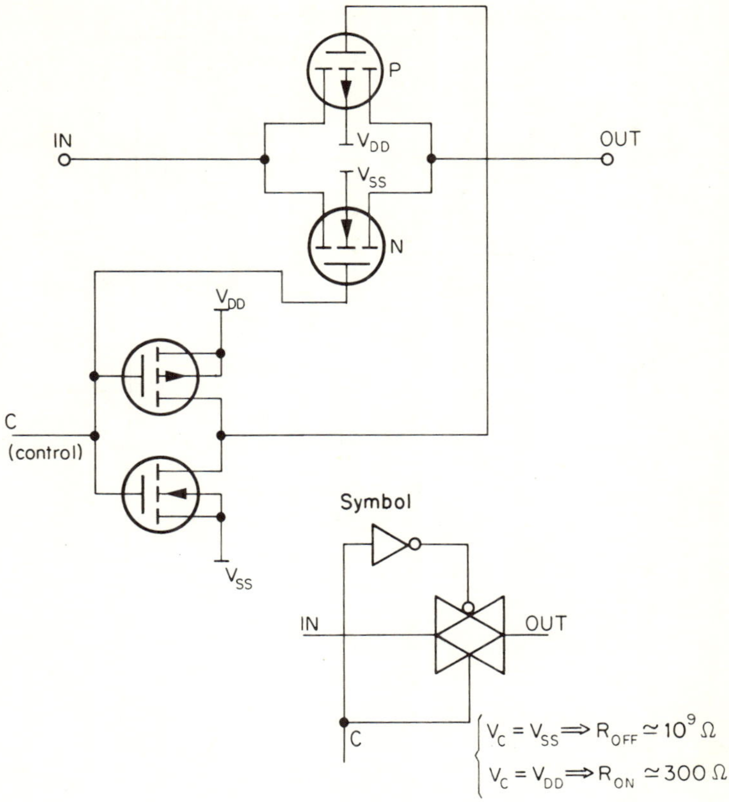

Figure 2.66

3c. Let us set *D* to the '1' state. The states of the flip-flop are modified as shown in Figure 2.72.

3d. With *D* held at '1', let the clock *C* become '1' also. The states of the flip-flop are modified as shown in Figure 2.73.

3e. When the clock *C* returns to '0', with input *D* held at '1', the states of the flip-flop do not change.

3f. With *D* retaking the '0' state and a clock pulse occurring, the same method allows the general timing diagram of Figure 2.74 to be obtained.

3g. This corresponds to the operating table of Figure 2.75, provided by the manufacturer.

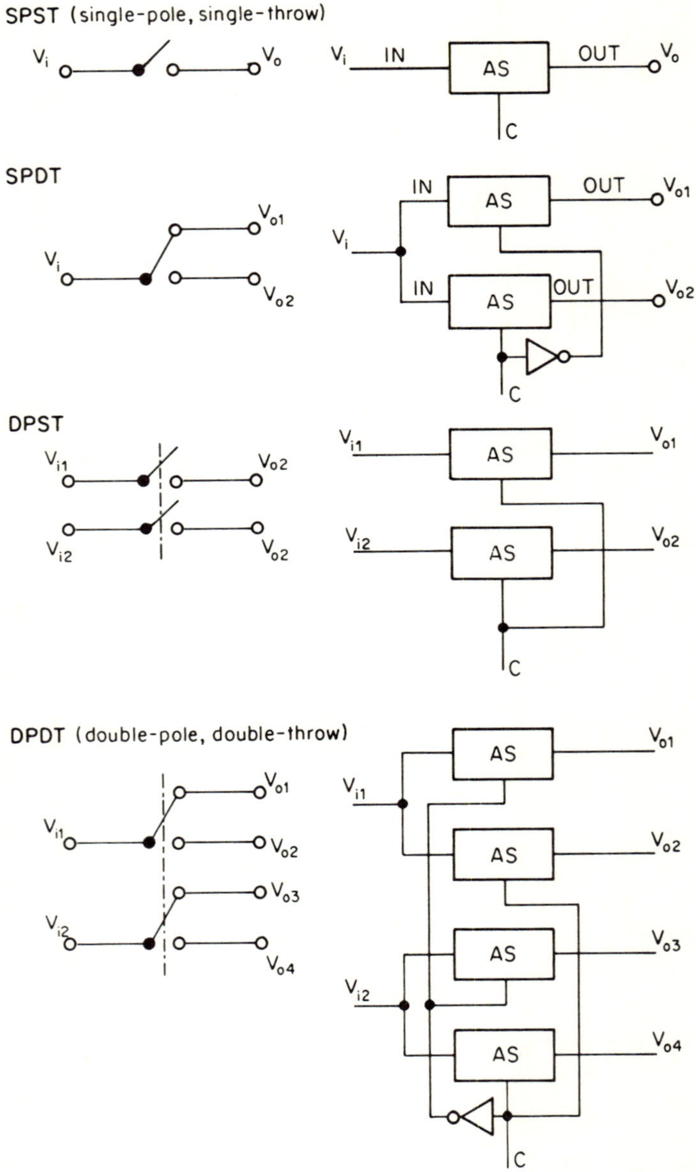
SPST (single-pole, single-throw)
V_i
V_o
IN
AS
OUT
C
SPDT
V_o1
V_o2
DPST
V_i1
V_i2
DPDT (double-pole, double-throw)
V_o3
V_o4

Figure 2.67

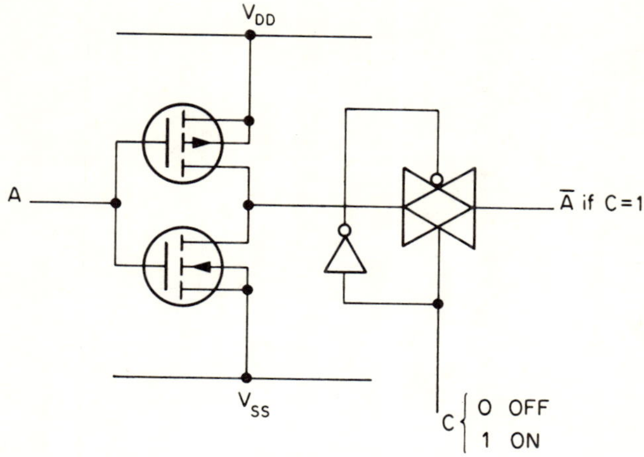

Figure 2.68

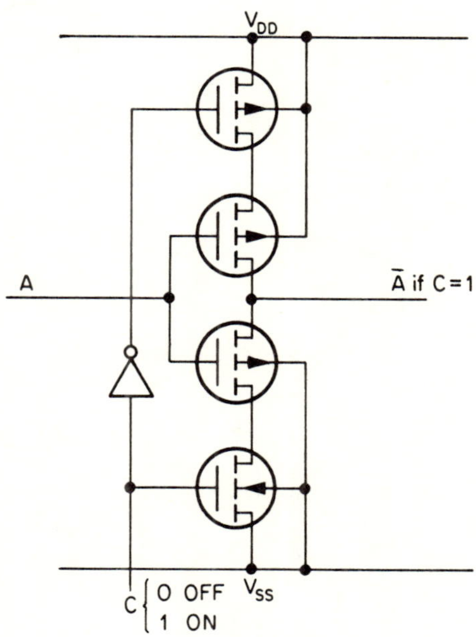

Figure 2.69

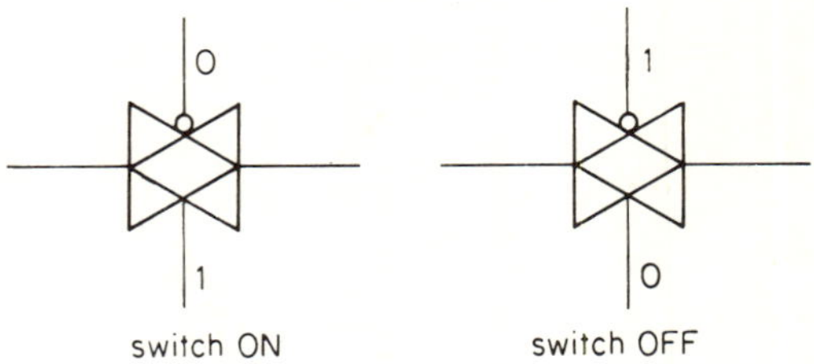

Figure 2.70

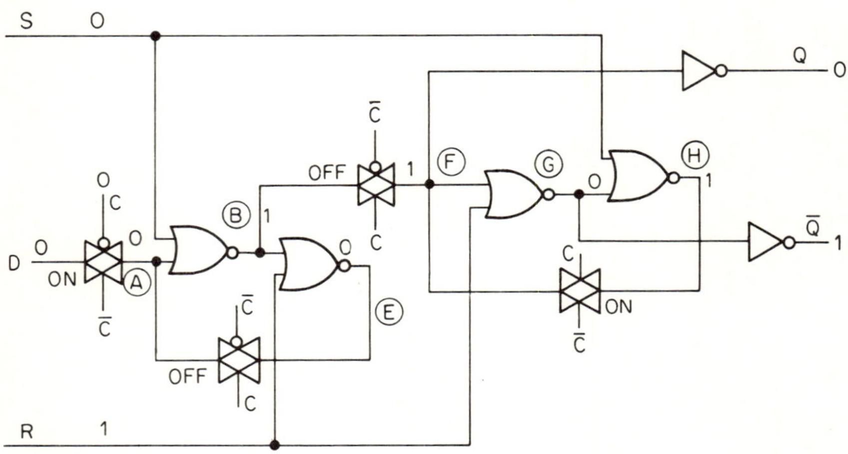

Figure 2.71

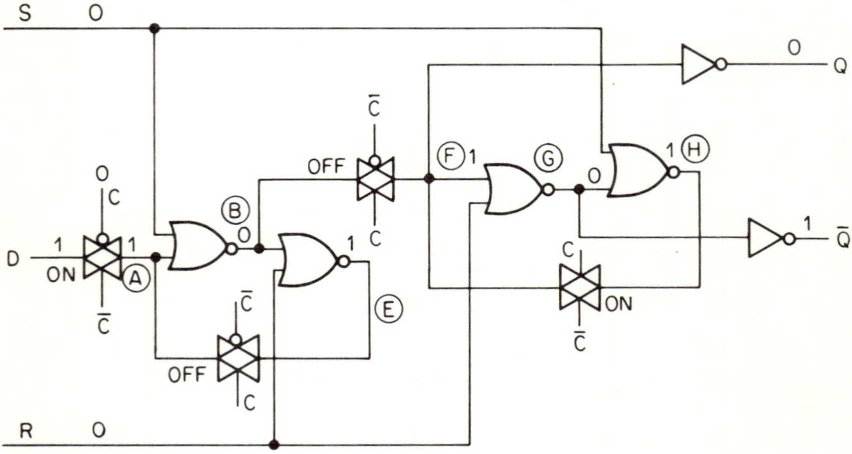

Figure 2.72

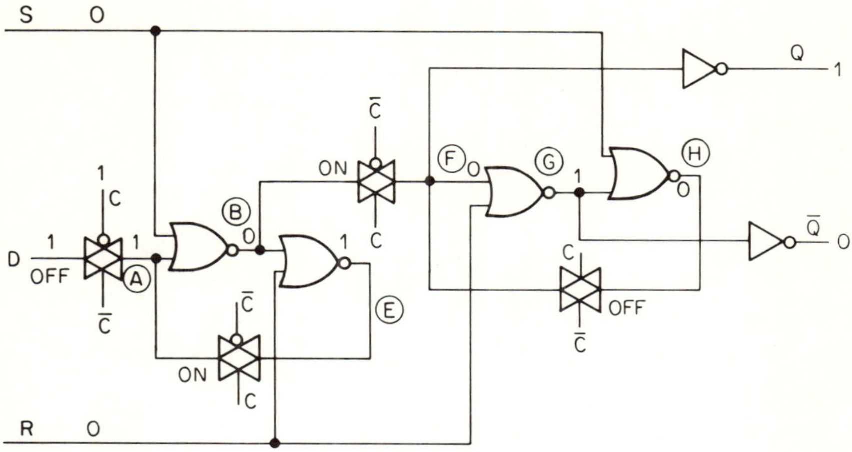

Figure 2.73

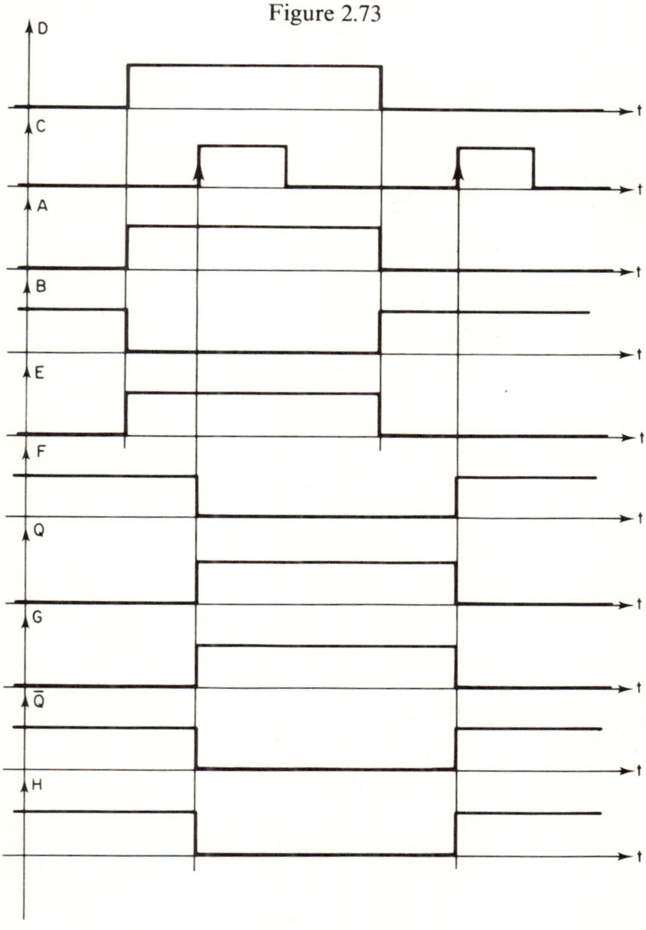

Figure 2.74

TRUTH TABLE

INPUTS				OUTPUTS		
CLOCK*	DATA	RESET	SET	Q	$\bar{Q}$	
	0	0	0	0	1	
	1	0	0	1	0	
	×	0	0	Q	$\bar{Q}$	No change
×	×	1	0	0	1	
×	×	0	1	1	0	
×	×	1	1	1	1	

× = Don't care
* = Level change

Figure 2.75

3 Combinational Systems

3.1 SUMMARY OF THEORY

A combinational system is characterized by the fact that a given combination of input variables has one and only one corresponding combination of output variables which appears after a time interval called the *propagation delay*. This is illustrated in Figure 3.1.

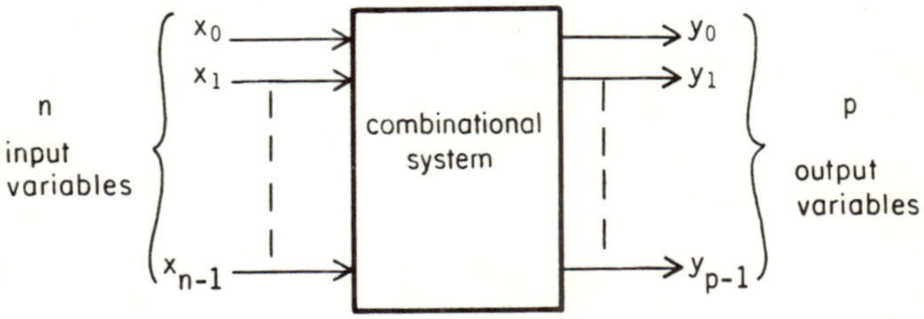

Figure 3.1

We have, in succession:

$$y_0 = f_0(x_0, x_1, \ldots, x_{n-1}),$$
$$y_1 = f_1(x_0, x_1, \ldots, x_{n-1}),$$
$$\vdots$$
$$y_{p-1} = f_{p-1}(x_0, x_1, \ldots, x_{n-1}),$$

from which can be written:

$$\mathbf{Y} = F(\mathbf{X}), \quad \text{where} \quad \mathbf{X} = (x_0, x_1, \ldots, x_{n-1}) \quad \text{and} \quad \mathbf{Y} = (y_0, y_1, \ldots, y_{p-1}).$$

X is called the *input vector* and **Y** the *output vector* (Figure 3.2).

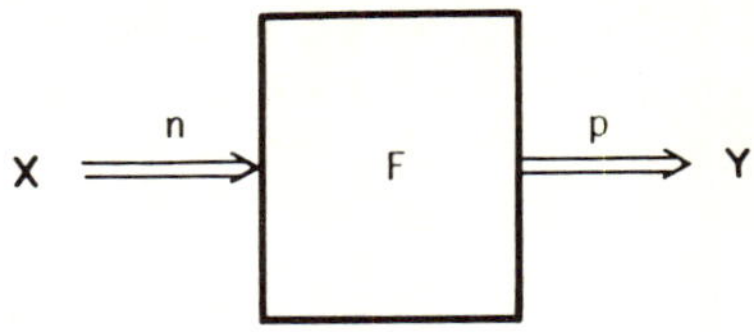

Figure 3.2

3.2 STATEMENT OF EXERCISES AND PROBLEMS

3.E.1 Parallel binary addition with serial carry propagation

The addition of two binary digits a and b is defined as shown in Figure 3.3. S is the sum and C is the carry to be *propagated.*

b	a	S	C
0	0	0	0
0	1	1	0
1	0	1	0
1	1	0	1

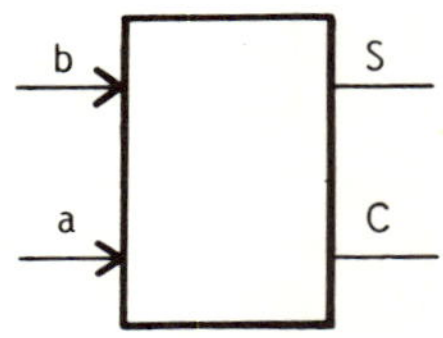

Figure 3.3

1. Deduce expressions for S and C.

2. Actually, the sum of any *number of columns* of binary digits must take account of the carry from the previous column:

$$\begin{array}{ccc} C_{n+1} & C_n & C_{n-1} \\ \ldots b_{n+1} & b_n & b_{n-1}\ldots \\ \ldots a_{n+1} & a_n & a_{n-1}\ldots \\ \hline & S_n & \end{array}$$

Complete the table of Figure 3.4, then deduce S_n and C_{n+1}.

c_n	b_n	a_n	S_n	C_{n+1}
0	0	0		
0	0	1		
0	1	0		
0	1	1		
1	0	0		
1	0	1		
1	1	0		
1	1	1		

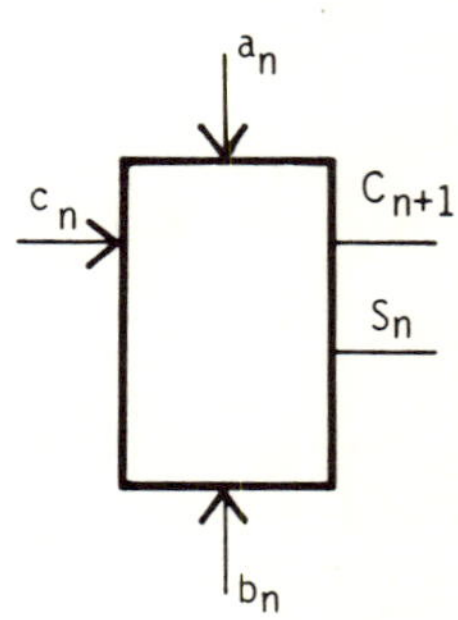

Figure 3.4

3. Synthesize this *full adder* using logic circuits from the TTL family.

4. Give the block diagram of a binary adder for 4-bit numbers, using the full adders modelled in Figure 3.4.

Here is an example of the addition of two 4-bit numbers:

$$\begin{array}{rrrl} & 0\leftarrow & \text{initial carry } C_0 & \\ & 1011 & 11 & \\ & +\,1101 & +\,13 & \\ \hline \text{final carry } C_4\rightarrow 1 & \underbrace{1000}_{\text{sum}} & 24 & \text{in base ten} \end{array}$$

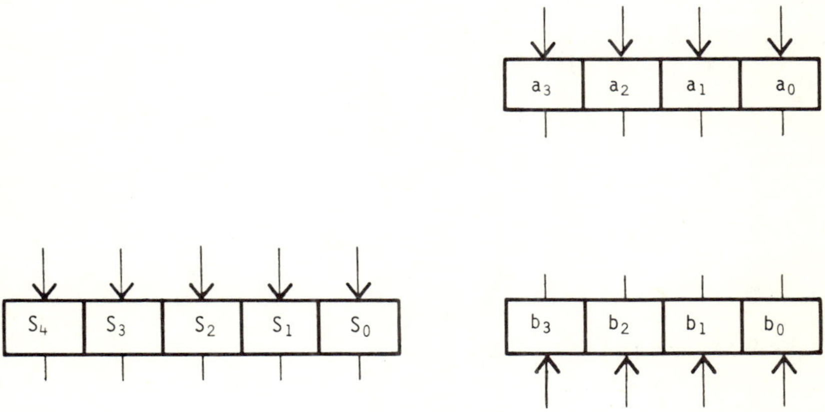

Figure 3.5

5. In practice, the result does not appear instantaneously. Each addition requires a time interval θ (ns) to calculate S_n and τ (ns) to calculate C_{n+1}.

Knowing that the propogation time of each constituent gate of the adder is 15 ns, calculate θ, τ, and the total calculation time T. Generalize if n is the number of stages.

Figure 3.6 gives the diagram of an adder of this type from 74 series TTL: the 74LS83.

3.E.2 Unsigned binary multiplication–combinational binary multiplication

3.E.2.1 One bit

1a. Knowing that *a* and *b* are two binary digits, define the arithmetic product of these two digits.

1b. Considering this table as a truth table for a logic function, what is the function?

1c. What is its symbol?

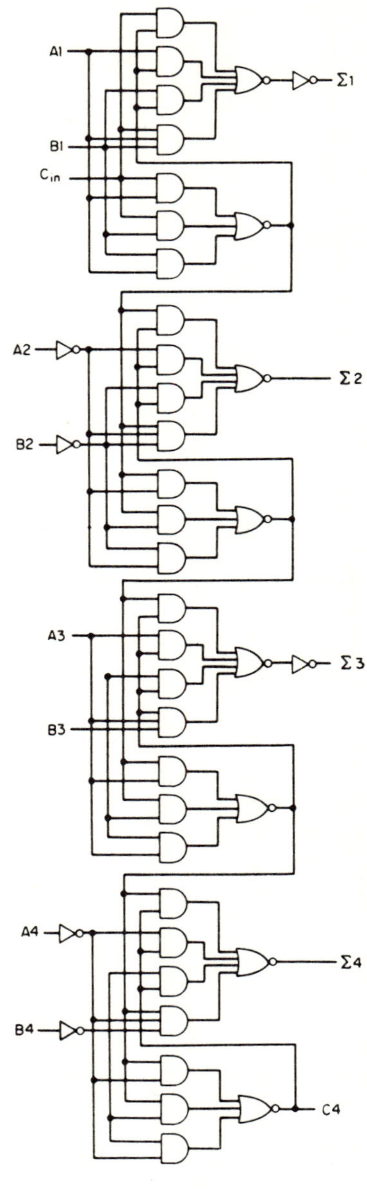

Figure 3.6

3.E.2.2 *Two bits*

Consider now two numbers of 2 bits: $A = a_2a_1$ and $B = b_2b_1$. The subscripts correspond to the weight of each variable.

We wish to realise $p = B \times A$ (Figure 3.7.)

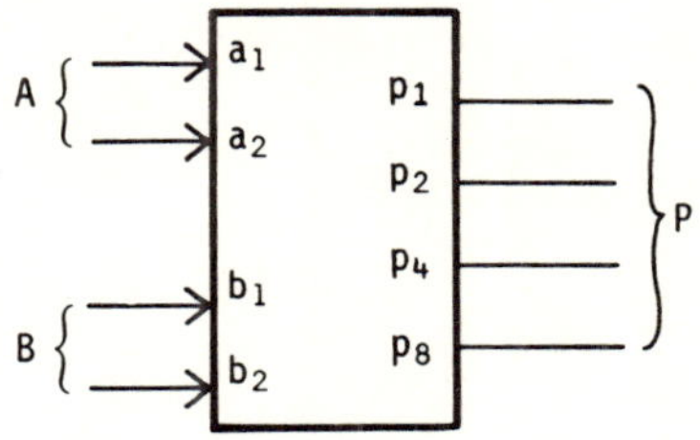

Figure 3.7

2a. Show that the product obtained is coded by 4 bits: p_1,p_2,p_4, and p_8. Consider n bits.

2b. Complete the table of Figure 3.8.

B		A		P			
b_2	b_1	a_2	a_1	p_8	p_4	p_2	p_1
0	0	0	0				
0	0	0	1				
0	0	1	0				
0	0	1	1				
0	1	0	0				
0	1	0	1				
0	1	1	0				
0	1	1	1				
1	0	0	0				
1	0	0	1				
1	0	1	0				
1	0	1	1				
1	1	0	0				
1	1	0	1				
1	1	1	0				
1	1	1	1				

Figure 3.8

2c. Express $p_i = f(b_2,b_1,a_2,a_1)$, for $i = 1,2,4$, and 8.

2d. Give the diagram of a multiplier by completing Figure 3.9.

2e. Obtain the same result by considering the operation performed manually.

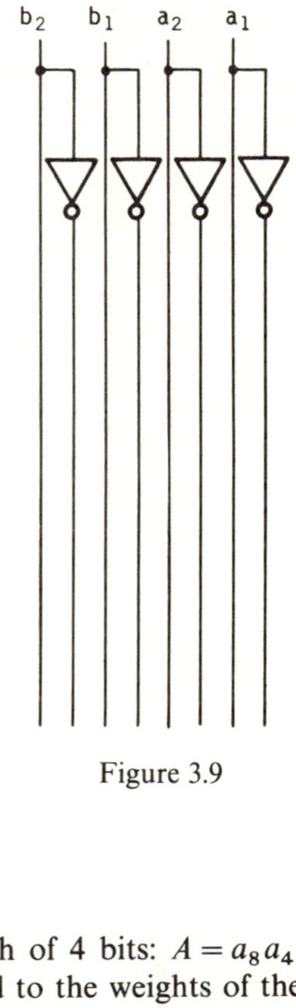

Figure 3.9

3.E.2.3 Four bits

Consider two numbers each of 4 bits: $A = a_8 a_4 a_2 a_1$ and $B = b_8 b_4 b_2 b_1$. The subscripts again correspond to the weights of the variables.

To realize $p = B \times A$, we could use the previous method; however, this leads to expressions for 8 functions of 8 variables. In order to avoid this difficulty, several solutions are possible.

3a. We shall consider an arrangement of 2×2 bit multipliers as studied in Section 3.E.2.2. Show that by using four multipliers of this type and some adders, it is possible to realize 4×4 bit multiplication. Give the corresonding diagram. Generalize to *n* bits.

3b. We can also use a matrix method, by first generating the 16 partial products $p_{ij} = a_i \times b_j$ (Figure 3.10).

Show that it is possible to obtain $p = A \times B$ by judicious addition of these 16 partial products.

Give a diagram of such a multiplier. Generalize to *n* bits.

3.E.3 Digital waveform generator

Consider the diagram of Figure 3.11, which can be symbolized by Figure 3.12. R_7 is not used.

The NAND gates of subcircuit 2 (Figure 3.11) have seven inputs, of which some are not used (connected to '1').

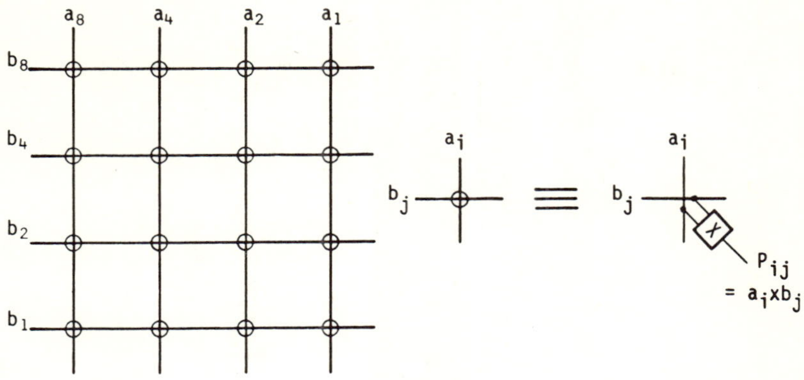

Figure 3.10

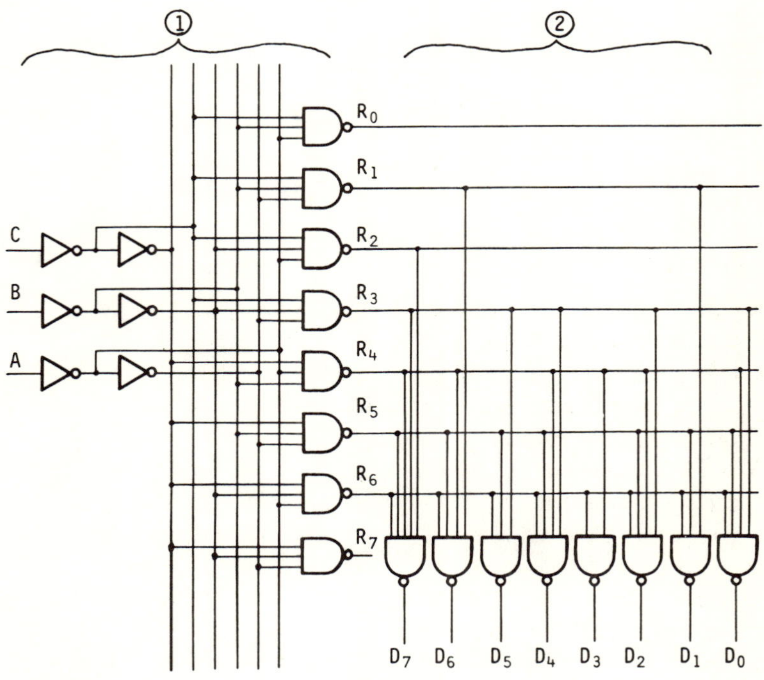

Figure 3.11

1. Express $R_i = f(C,B,A)$, for $i = 0$ to 6.
2. What name can be given to subscript 1?
3. Complete the table of Figure 3.13.
4. Express $D_i = f(R_i)$, for $i = 0$ to 7.
5. Express $D_i = f(C,B,A)$, without simplification, in the form of a sum of products.

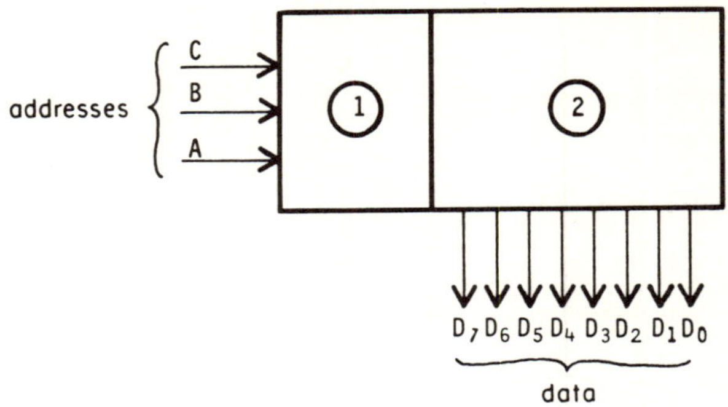

Figure 3.12

	4	2	1															
Addr	C	B	A	R_0	R_1	R_2	R_3	R_4	R_5	R_6	D_7	D_6	D_5	D_4	D_3	D_2	D_1	D_0
0	0	0	0															
1	0	0	1															
2	0	1	0															
3	0	1	1															
4	1	0	0															
5	1	0	1															
6	1	1	0															
7	1	1	1			not used												

Figure 3.13

6. Simplify the expressions obtained.
7. Draw the corresponding diagram using NAND gates.
8. Compare with the diagram of Figure 3.11; what can be deduced?
9. Subcircuits 1 and 2, without simplification, represent a *read only memory* (ROM), since at each address information is stored which will now be used.

The outputs D_i are followed by an interface which combines V_{ei} and D_i such that

$$V_{ei} = 0\,\text{V} \quad \text{if } D_i = \text{logic } 0,$$
$$V_{ei} = 10\,\text{V} \quad \text{if } D_i = \text{logic } 1.$$

In their turn, the outputs V_{ei} control the summing circuit of Figure 3.14.

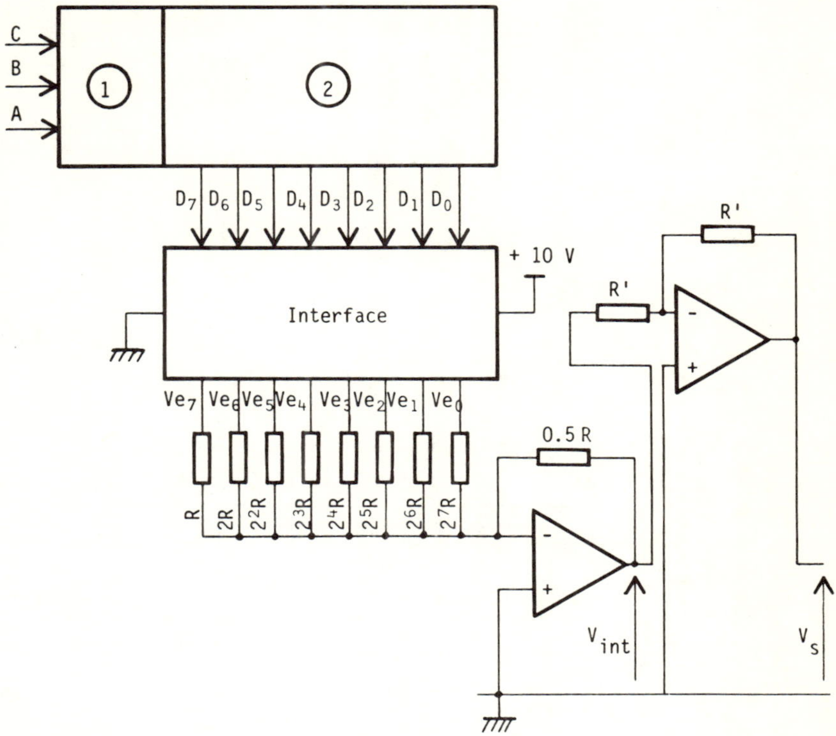

Figure 3.14

We shall apply the inputs CBA, in a periodic manner, with the following address seqence:

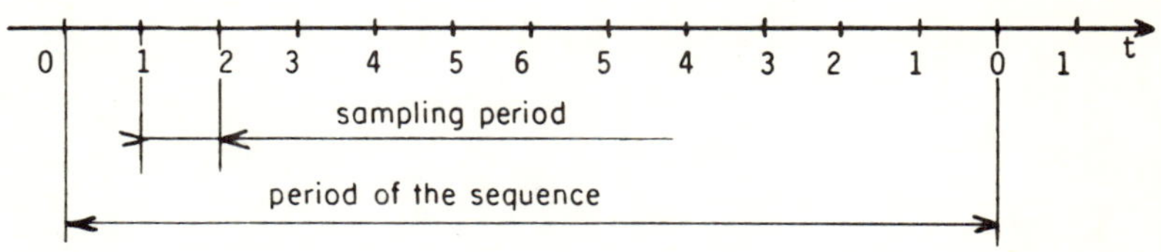

$$\text{Express} \quad V_S = f(D_7, D_6, \ldots, D_0).$$

What name can be given to this arrangement?

10. Calculate V_S for addresses 0,1,...,6 by completing the table of Figure 3.15.

11. Trace the curve $V_S = f(t)$ by completing the graph of Figure 3.16.

12. Following Figure 3.16 and by completing the table of Figure 3.17, show how a sinusoidal voltage could be obtained.

13. Give the block diagram of a complete arrangement which allows the required waveform to be generated and hence, a detailed circuit.

Note: The angles α are between 0 and 90° and have integral values.

14. How could the process be extended to any waveform?

Addr	D_7	D_6	D_5	D_4	D_3	D_2	D_1	D_0	V_S
0									
1									
2									
3									
4									
5									
6									

Figure 3.15

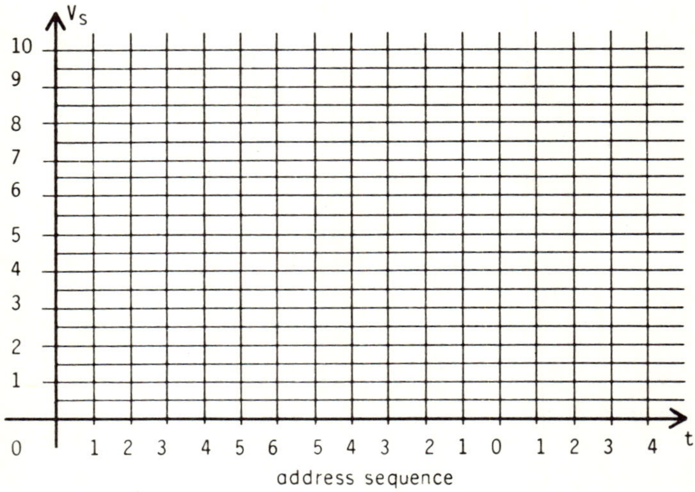

Figure 3.16

Addr	V_S	$V_S/10 = \sin\alpha$	α (°)
0			
1			
2			
3			
4			
5			
6			

Figure 3.17

3.3 SOLUTIONS

3.S.1 Parallel binary addition with serial carry propagation

1.
$$S = a \oplus b$$
$$C = a \cdot b$$

2. The complete table is given in Figure 3.18.

c_n	b_n	a_n	S_n	C_{n+1}
0	0	0	0	0
0	0	1	1	0
0	1	0	1	0
0	1	1	0	1
1	0	0	1	0
1	0	1	0	1
1	1	0	0	1
1	1	1	1	1

$S_n = a_n \oplus b_n \oplus c_n$

$C_{n+1} = a_n \cdot b_n + c_n \cdot (a_n + b_n)$

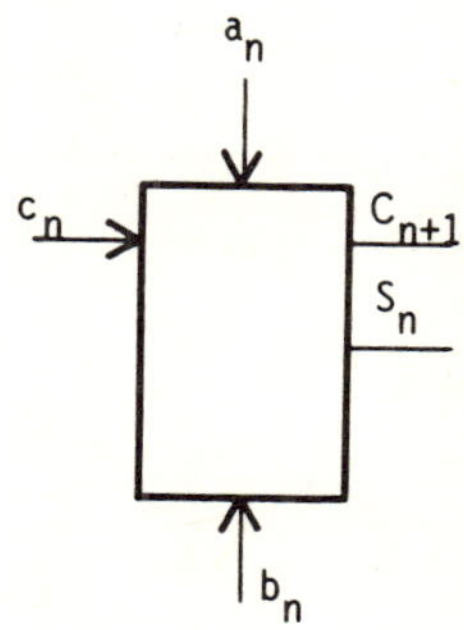

Figure 3.18

3. The diagram of a full adder is given in Figure 3.19.

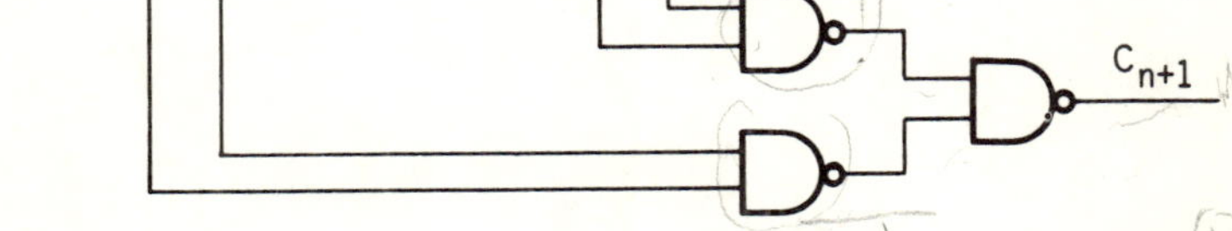

Figure 3.19

It is obtained by noticing that the carry can also be written:

$$C_{n+1} = a_n \cdot b_n + c_n \cdot (a_n \oplus b_n),$$

hence
$$C_{n+1} = (a_n / b_n)/[c_n/(a_n \oplus b_n)].$$

4. The diagram of a parallel 4-bit adder is given in Figure 3.20.

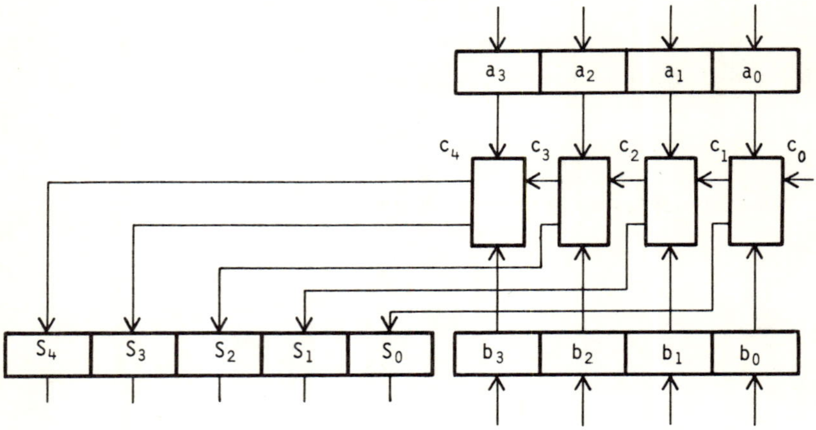

Figure 3.20

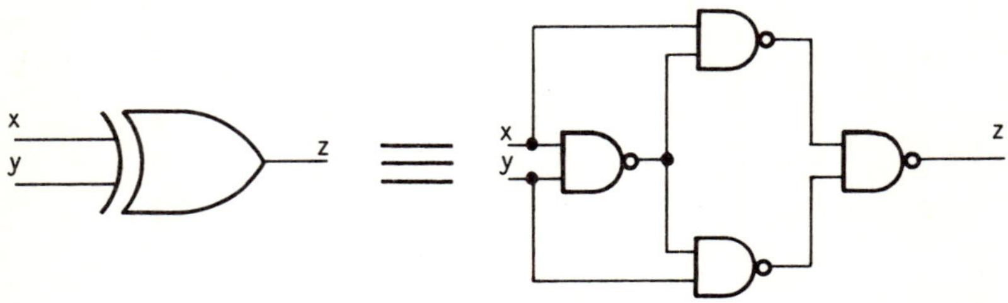

Figure 3.21

5. In practice, each exclusive OR circuit corresponds to Figure 3.21, hence a propagation time of 45 ns. So the complete addition requires (Figure 3.22):

$$\theta = 45 \times 2 = 90\,\text{ns},$$
$$\tau = 45 + (15 \times 2) = 75\,\text{ns}.$$

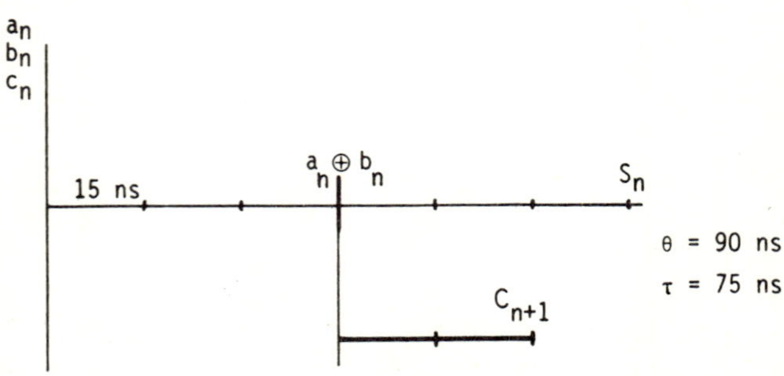

Figure 3.22

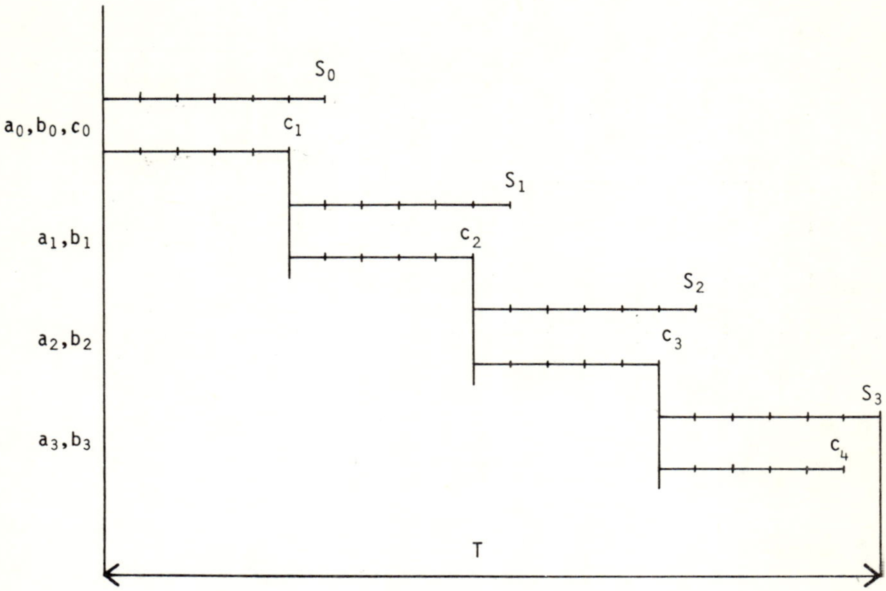

Figure 3.23

The total calculation time T will be obtained from Figure 3.23,
hence $T = \tau \times 3 + \theta = 75 \times 3 + 90 = 315$ ns.
In general: $T = \tau(n-1) + \theta$

3.S.2 Unsigned binary multiplication

3.S.2.1 One bit

It is easy to obtain the following results:

b	a	P=axb
0	0	0
0	1	0
1	0	0
1	1	1

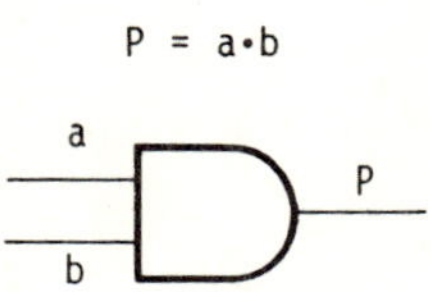

3.S.2.2 Two bits

2a. With n bits, the numbers 0 to $2^n - 1$ can be represented. The highest value reached by the product is, therefore, $(2^n - 1)^2 = 2^{2n} - 2^{n+1} + 1$. Representing this number:

rank	→	2n	(2n-1)	...	(n+2)	(n+1)	n	(n-1)	...1	0
$2^{2n}+1$	→	1	0	...	0	0	0	0	...0	1
2^{n+1}	→	0	0	...	0	1	0	0	...0	0
$2^{2n}+1-2^{n+1}$	→	0	1	...	1	1	0	0	...0	1

(n bits: ranks (n-1)...0; 2n bits: ranks (2n-1)...0)

Therefore the result requires $2n$ bits. In this case 4.

2b. The truth table is given in Figure 3.24.

2c. Four K maps allow us to write:

$$p_1 = a_1 \cdot b_1$$

$$p_2 = \bar{b}_2 \cdot b_1 \cdot a_2 + b_2 \cdot \bar{b}_1 \cdot a_1 + b_2 \cdot \bar{a}_2 \cdot a_1 + b_1 \cdot a_2 \cdot \bar{a}_1$$

$$= a_2 \cdot b_1 (\overline{a_1} + \overline{b_2}) + a_1 \cdot b_2 \cdot (\overline{a_2} + \overline{b_1})$$

$$= a_2 \cdot b_1 \cdot \overline{a_1 \cdot b_2} + a_1 \cdot b_2 \cdot \overline{a_2 \cdot b_1}$$

$$= (a_2 \cdot b_1) + (a_1 \cdot b_2)$$

$$P_4 = b_2 \cdot \overline{b_1} \cdot a_2 + b_2 \cdot a_2 \cdot \overline{a_1} = a_2 \cdot b_2 \cdot \overline{a_1 \cdot b_1}$$

$$p_8 = a_2 \cdot b_2 \cdot a_1 \cdot b_1$$

B		A		P			
b_2	b_1	a_2	a_1	p_8	p_4	p_2	p_1
0	0	0	0	0	0	0	0
0	0	0	1	0	0	0	0
0	0	1	0	0	0	0	0
0	0	1	1	0	0	0	0
0	1	0	0	0	0	0	0
0	1	0	1	0	0	0	1
0	1	1	0	0	0	1	0
0	1	1	1	0	0	1	1
1	0	0	0	0	0	0	0
1	0	0	1	0	0	1	0
1	0	1	0	0	1	0	0
1	0	1	1	0	1	1	0
1	1	0	0	0	0	0	0
1	1	0	1	0	0	1	1
1	1	1	0	0	1	1	0
1	1	1	1	1	0	0	1

B \ A	0	1	2	3
0	0	0	0	0
1	0	1	2	3
2	0	2	4	6
3	0	3	6	9

Figure 3.24

2d. The diagram is given in Figure 3.25.

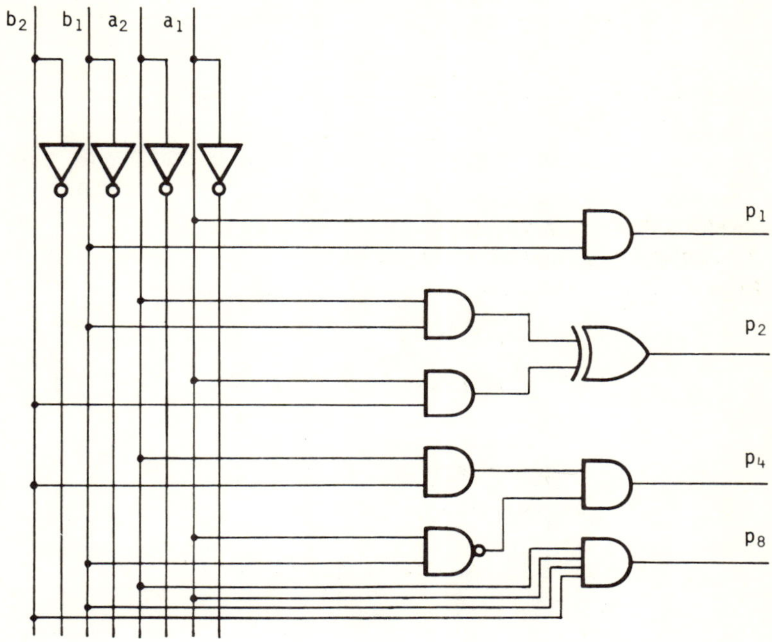

Figure 3.25

2e. Consider the operation:

weights →	8	4	2	1
			a_2	a_1
			$\times\, b_2$	b_1
	C_8 ←	C_4 ←	$a_2\cdot b_1$	$a_1\cdot b_1$
	+	$a_2\cdot b_2$	$a_1\cdot b_2$	
	p_8	p_4	p_2	p_1

$$p_1 = a_1\cdot b_1$$

$$p_2 = (a_2\cdot b_1)\oplus(a_1\cdot b_2) \qquad \text{and} \qquad C_4 = (a_2\cdot b_1)\cdot(a_1\cdot b_2)$$

results from the addition of 2 bits (see problems 3.E.1)

$$p_4 = (a_2\cdot b_2)\oplus(a_2\cdot b_1\cdot a_1\cdot b_2) = a_2\cdot b_2\overline{a_1\cdot b_1}$$

$$p_8 = C_8 = (a_2\cdot b_2)\cdot(a_2\cdot b_1\cdot a_1\cdot b_2) = a_2\cdot b_2\cdot a_1\cdot b_1$$

3.S.2.3. Four bits

3a. Consider $A_4 = a_8 a_4$, $A_1 = a_2 a_1$

$B_4 = b_8 b_4$, $B_1 = b_2 b_1$

We can write $A = 4A_4 + A_1$

$B = 4B_4 + B_1$

It follows that $p = B \times A = 16 B_4 \times A_4 + 4(B_4 \times A_1 + B_1 \times A_4) + B_1 \times A_1$

The products $B_4 \times A_4, B_4 \times A_1, B_1 \times A_4$, and $B_1 \times A_1$ correspond to the outputs of four 2 × 2 bit multipliers of the type studied peviously. The expression for P shows that the sum of these products can be obtained as shown in Figure 3.26.

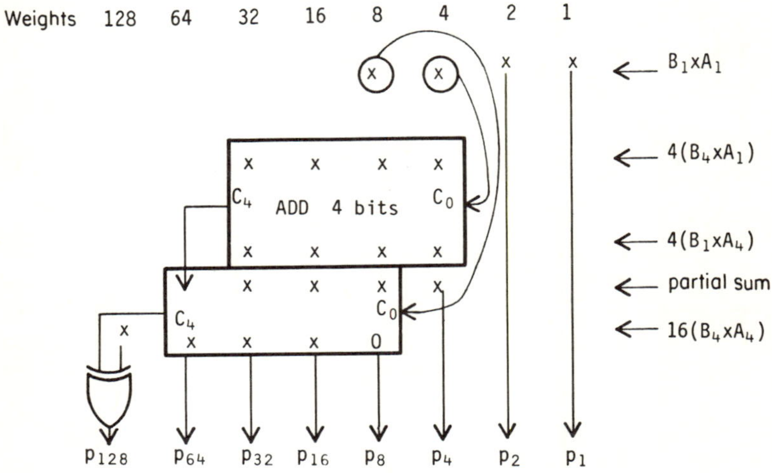

Figure 3.26

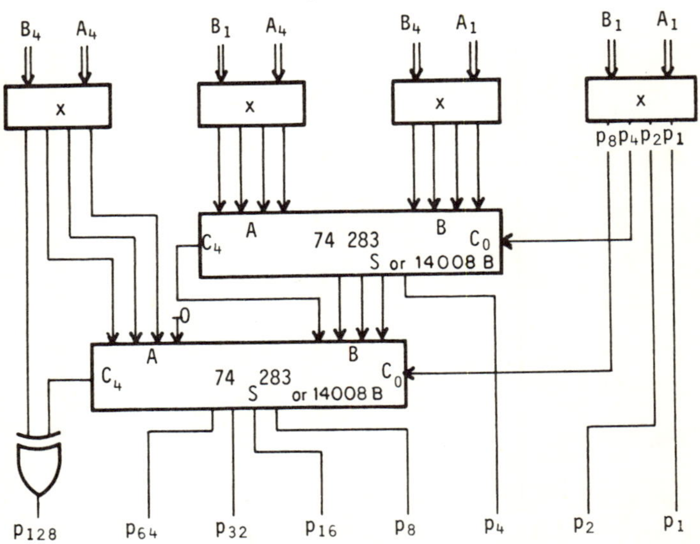

Figure 3.27

The corresponding diagram is given in Figure 3.27.
In the general case of $n \times n$ bit multiplication, $p \times p$ bit multipliers are arranged so that $2p = n$, and the previous scheme is not exactly generalizable. See Figure 3.28.

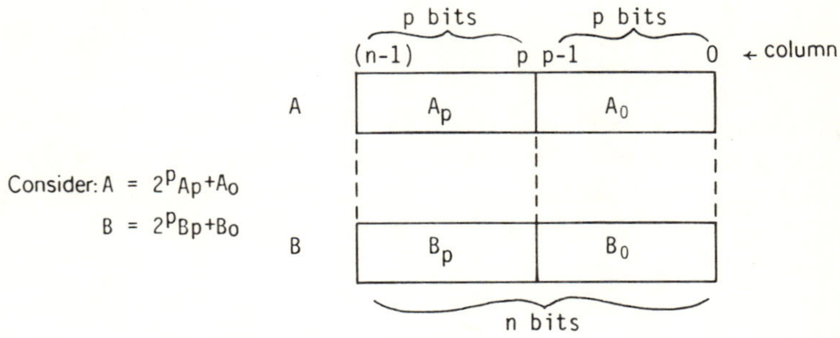

Figure 3.28

It follows that:

$$P = B \times A = 2^{2p} B_p \times A_p + 2^p (B_p \times A_0 + B_0 \times A_p) + B_0 \times A_0.$$

The products $B_p \times A_p$, $B_p \times A_0$, $B_0 \times A_p$, $B_0 \times A_0$, correspond to the outputs of four $p \times p$-bit multipliers. The expression for P shows that the summation of these products can be realized as shown in Figure 3.29.

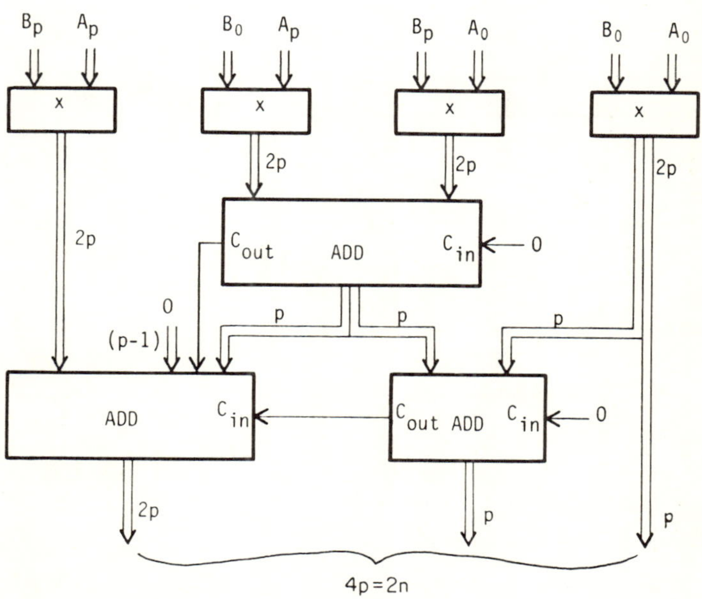

Figure 3.29

It is easy to assess the components required: four $p \times p$-bit multipliers, two $2p$-bit adders, and one p-bit adder.

3b. The product may be written:

$$P = \underbrace{(8a_8 + 4a_4 + 2a_2 + a_1)}_{A} + \underbrace{(8b_8 + 4b_4 + 2b_2 + b_1)}_{B}$$

$$\text{or } P = 8b_8 \cdot A + 4b_4 \cdot A + 2b_2 \cdot A + b_1 \cdot A.$$

$b_8 \cdot A$, $b_4 \cdot A$, $b_2 \cdot A$, and $b_1 \cdot A$ represent the rows of the matrix of Figure 3.10. This expression for P shows that it is sufficient to add these rows with an offset for each column. Consequently, after each addition, the rightmost bit is one bit of the result. This is sumarized in Figure 3.30.

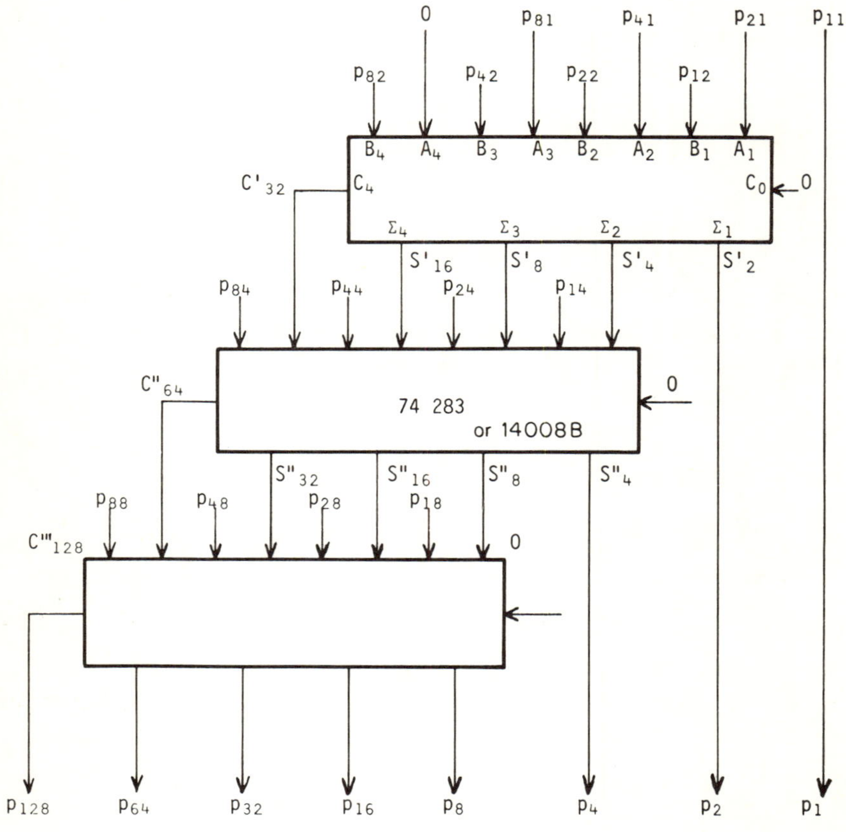

Figure 3.30

Generalization to n bits is immediate: to multiply n bits by n bits, n^2 2-input AND gates and $(n-1)$ n-bit adders are required.

The calculation time is:

$$T = t \quad \text{propag. AND} + (n-1) \times t \quad \text{propag. ADD}$$

3.S.3 Digital waveform generator

1. The expressions for R_i are as follows:

$$R_0 = \overline{\bar{C} \cdot \bar{B} \cdot \bar{A}} \qquad R_1 = \overline{\bar{C} \cdot \bar{B} \cdot A} \qquad R_2 = \overline{\bar{C} \cdot B \cdot \bar{A}} \qquad R_3 = \overline{\bar{C} \cdot B \cdot A}$$

$$R_4 = \overline{C \cdot \bar{B} \cdot \bar{A}} \qquad R_5 = \overline{C \cdot \bar{B} \cdot A} \qquad R_6 = \overline{C \cdot B \cdot \bar{A}}$$

2. The subcircuit one is a decoder of the input variables and hence the addresses.

3. The completed table is given in Figure 3.31.

Addr	C	B	A	R_0	R_1	R_2	R_3	R_4	R_5	R_6	D_7	D_6	D_5	D_4	D_3	D_2	D_1	D_0
0	0	0	0	0	1	1	1	1	1	1	0	0	0	0	0	0	0	0
1	0	0	1	1	0	1	1	1	1	1	0	1	0	0	0	0	1	0
2	0	1	0	1	1	0	1	1	1	1	1	0	0	0	0	0	0	0
3	0	1	1	1	1	1	0	1	1	1	1	0	1	1	0	1	0	1
4	1	0	0	1	1	1	1	0	1	1	1	1	0	1	1	1	0	1
5	1	0	1	1	1	1	1	1	0	1	1	1	1	1	0	1	1	1
6	1	1	0	1	1	1	1	1	1	0	1	1	1	1	1	1	1	1
7	1	1	1	not used														

Figure 3.31

4. The expressions for D_i are as follows:

$$D_0 = \overline{R_3 \cdot R_4 \cdot R_5 \cdot R_6} \quad D_1 = \overline{R_1 \cdot R_5 \cdot R_6} \quad D_2 = D_0 \quad D_3 = \overline{R_4 \cdot R_6}$$

$$D_4 = D_0 \quad D_5 = \overline{R_3 \cdot R_5 \cdot R_6} \quad D_6 = \overline{R_1 \cdot R_4 \cdot R_5 \cdot R_6} \quad D_7 = \overline{R_2 \cdot R_3 \cdot R_4 \cdot R_5 \cdot R_6}$$

5. The expressions for D_i are obtained by replacing the R_i in the expressions obtained in 4 by their values obtained in 1 and simplifying:

$$D_0 = D_2 = D_4 = \bar{C} \cdot B \cdot A + C \cdot \bar{B} \cdot \bar{A} + C \cdot \bar{B} \cdot \bar{A} + C \cdot B \cdot \bar{A}$$
$$D_1 = \bar{C} \cdot \bar{B} \cdot A + C \cdot \bar{B} \cdot A + C \cdot B \cdot \bar{A}$$
$$D_3 = C \cdot \bar{B} \cdot \bar{A} + C \cdot B \cdot \bar{A}$$
$$D_5 = \bar{C} \cdot B \cdot A + C \cdot \bar{B} \cdot A + C \cdot B \cdot \bar{A}$$
$$D_6 = \bar{C} \cdot \bar{B} \cdot A + C \cdot \bar{B} \cdot \bar{A} + C \cdot \bar{B} \cdot A + C \cdot B \cdot \bar{A}$$
$$D_7 = \bar{C} \cdot B \cdot \bar{A} + \bar{C} \cdot B \cdot A + C \cdot \bar{B} \cdot \bar{A} + C \cdot \bar{B} \cdot A + C \cdot B \cdot \bar{A}.$$

6. Figure 3.32 gives the eight K maps relating to these functions.

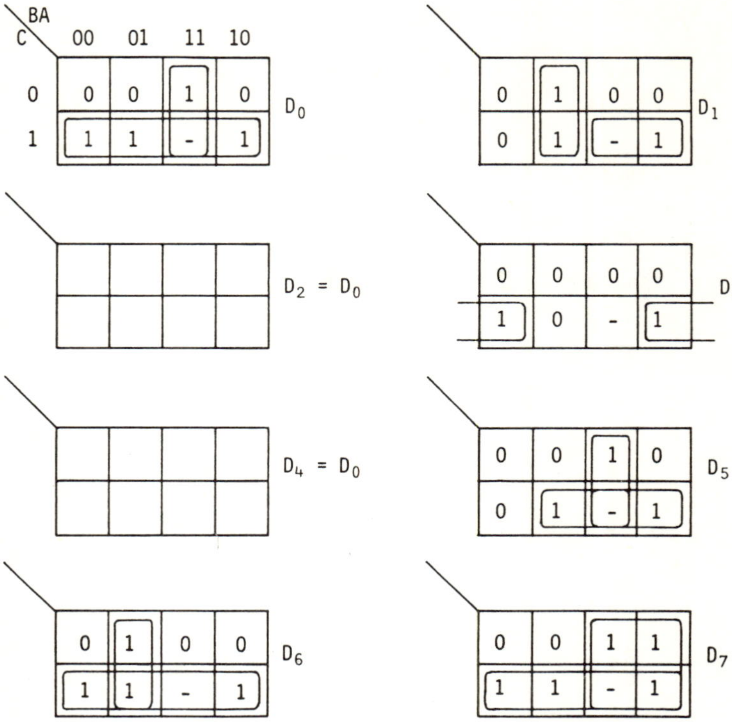

Figure 3.32

We obtain:

$$D_0 = D_2 = D_4 = C + B \cdot A = \bar{C}/(B/A)$$
$$D_1 = \bar{B} \cdot A + C \cdot B = (\bar{B}/A)/(C/B)$$
$$D_3 = C \cdot \bar{A} = (C/\bar{A})/1$$
$$D_5 = B \cdot A + C \cdot B + C \cdot A = (B/A)/(C/B)/(C/A)$$
$$D_6 = C + \bar{B} \cdot A = \bar{C}/(\bar{B}/A)$$
$$D_7 = C + B = \bar{C}/\bar{B}$$

7. The diagram is given in Figure 3.33.

8. The simplification which can be achieved is determined by the number of inputs and not by the number of circuits.

For the diagram of Figure 3.11, it is clear that information coded by the 8 variables D_7 to D_0 corresponds to each address present at the inputs CBA.

Each NAND of subcircuit 2 can be connected to 7 of the 8 variables R_7 to R_0. If a NAND was connected to 8 variables, its output would be always at '1', since one of the 8 variables R_7 to R_0 must be at '0'. In this way, by using eight 7-input NAND gates, it is possible to *change* the stored information, in other words to *program* the equivalent of a read only memory (ROM) (mentioned in problem 9).

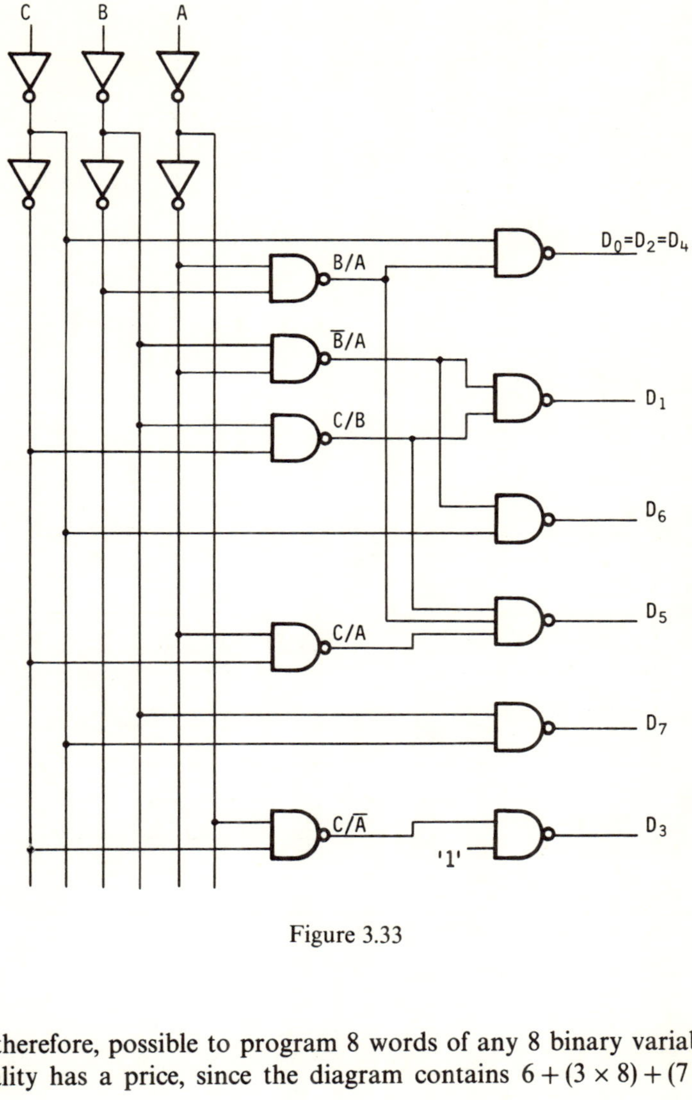

Figure 3.33

It is, therefore, possible to program 8 words of any 8 binary variables. This universality has a price, since the diagram contains $6+(3\times 8)+(7\times 8)=86$ inputs.

For the diagram of Figure 3.33, the number of inputs is $6+(2\times 10)+3=29$. However, this diagram corresponds to an application of ROM (considered below).

9. The second operational amplifier is used as an inverter of which the gain is $V_s/V_{\text{int}} = -R'/R' = -1$.

The first amplifier is used as an inverting summer of which the output voltage is:

$$V_{\text{int}} = -0.5\,R\left(\frac{V_{e7}}{R}+\frac{V_{e6}}{2R}+\ldots+\frac{V_{e0}}{128R}\right) \qquad (1)$$

Addr	D_0	D_1	D_2	D_3	D_4	D_5	D_6	D_7	V_s
0	0	0	0	0	0	0	0	0	0
1	0	1	0	0	0	0	1	0	2.58
2	1	0	0	0	0	0	0	0	5
3	1	0	1	1	0	1	0	1	7.07
4	1	1	0	1	1	1	0	1	8.63
5	1	1	1	1	0	1	1	1	9.65
6	1	1	1	1	1	1	1	1	9.96

Figure 3.34

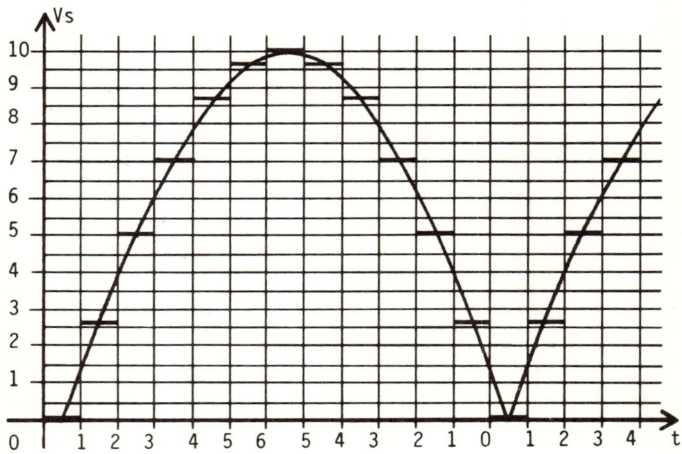

Figure 3.35

Addr	V_s	$V_s/10=\sin\alpha$	α
0	0	0	0°
1	2.58	0.258	15°
2	5	0.5	30°
3	7.07	0.705	45°
4	8.63	0.863	60°
5	9.65	0.965	75°
6	9.96	0.996	90°

Figure 3.36

Now it is possible to put $V_{ei} = 10\text{D}_i$. So we have:

$$V_{ei} = 10\,\text{V} \quad \text{if} \quad D_i = \text{'0'}$$
$$V_{ei} = 10\,\text{V} \quad \text{if} \quad D_i = \text{'1'}$$

Expression (1) becomes:

$$V_{\text{int}} = -\frac{10 \times 0.5R}{R}\left(D_7 + \frac{D_6}{2} + \frac{D_5}{4} + \ldots + \frac{D_0}{128}\right)$$
$$= -\tfrac{5}{128}(128D_7 + 64D_6 + 32D_5 + \ldots + D_0).$$

Hence:

$$V_s = -V_{\text{int}} = \tfrac{5}{128}(128D_7 + 64D_6 + \ldots + D_0). \qquad (2)$$

The interface and summer enable a voltage V_s to be associated with a digital value coded by the 8 binary variables D_7 to D_0; this arrangement is called a *digital to analog converter* (DAC).

10. The completed table of Figure 3.15 is given in Figure 3.34. Figure 3.31 gives the D_i as functions of the addresses; relation (2) may now be applied.

11. Figure 3.35 is obtained by entering the values obtained on the graph of Figure 3.16.

12. The completed table of Figure 3.17 is given in Figure 3.36.

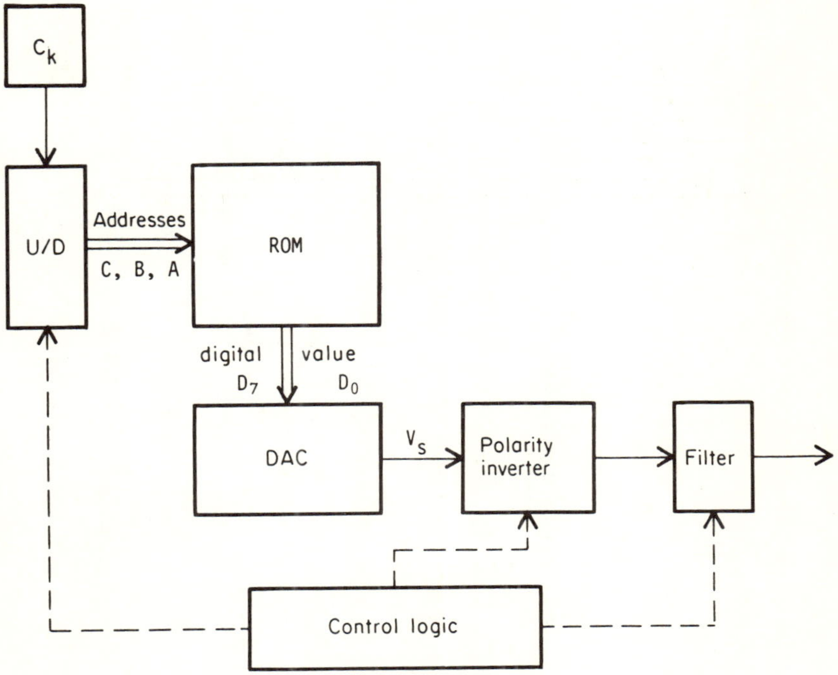

Figure 3.37

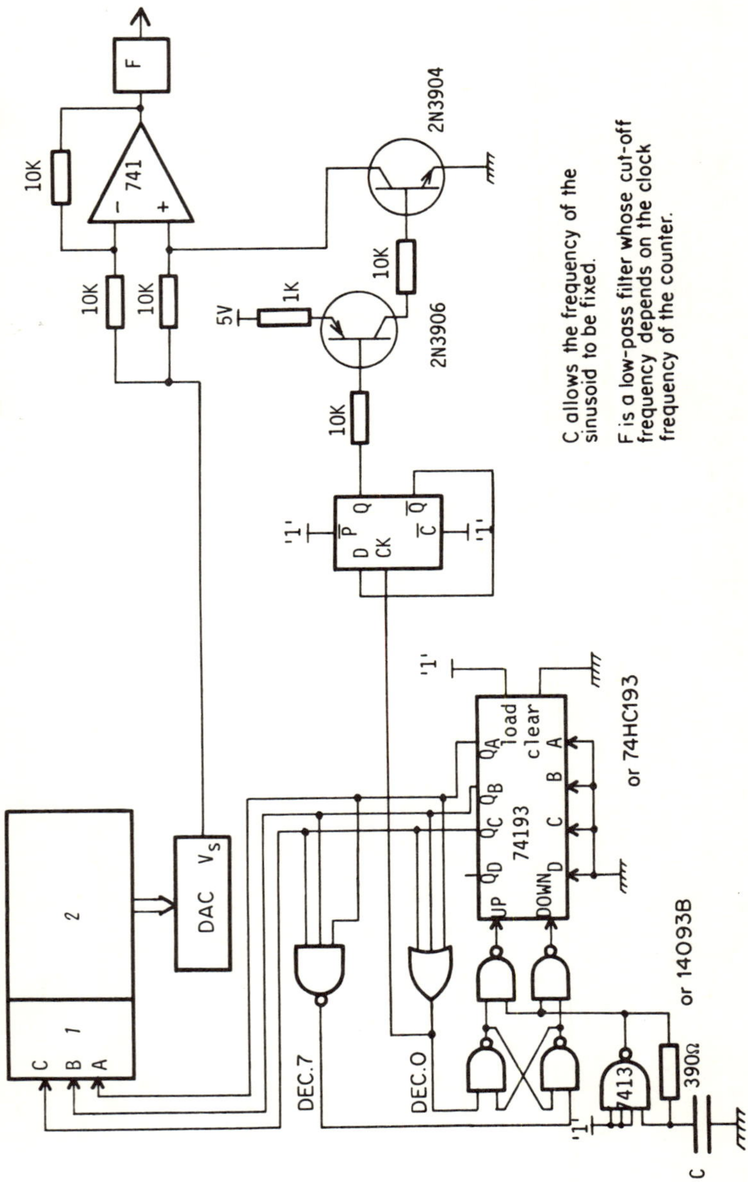
F
10K
741
10K
10K
5V
1K
2N3904
10K
2N3906
10K
'1'
D
CK
P
Q
C
Q
'1'
'1'
load
clear
74193
or 74HC193
UP
DOWN
2
DAC
V_S
C
B
A
1
DEC.7
DEC.0
or 14093B
7413
390Ω
C
C allows the frequency of the sinusoid to be fixed.
F is a low-pass filter whose cut-off frequency depends on the clock frequency of the counter.

Figure 3.38

13. The waveform obtained is actually a 'digital wave' in the form of a full-wave rectified sinusoid. In order to obtain a complete sinusoid it is necessary to:

(a) Use an up/down counter U/D driven by a clock Ck in order to generate the succession of addresses.
(b) Use a polarity inverter in order to create the negative half cycle.
(c) Use a filter in order to smooth the curve.

The block diagram of such an arrangement is given in Figure 3.37.

Control logic *manages* the arrangement. A detailed circuit, using classical TTL components, is given in Figure 3.38.

14. The procedure is easily extended to any waveform if the ROM can be programmed with the values corresponding to the required waveform. Certainly, the ROM used will be a user-programmable integrated circuit, called PROM or, better EPROM (which can also be erased).

4 *Sequential systems*

4.1 SUMMARY OF THEORY

A sequential system is characterized by the fact that several combinations of the output variables can correspond to one combination of the input variables; the combination which appears depends on the combination which preceded a change in one of the input variables. In other words, the present state depends closely on the preceding states: the circuit remembers its previous history. This is shown in Figure 4.1.

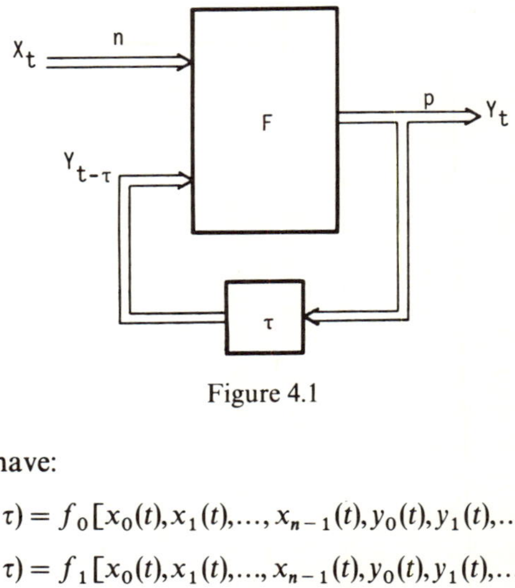

Figure 4.1

In succession we have:

$$y_0(t+\tau) = f_0[x_0(t), x_1(t), \dots, x_{n-1}(t), y_0(t), y_1(t), \dots, y_{p-1}(t)]$$
$$y_1(t+\tau) = f_1[x_0(t), x_1(t), \dots, x_{n-1}(t), y_0(t), y_1(t), \dots, y_{p-1}(t)]$$
$$\vdots$$
$$y_{p-1}(t+\tau) = f_{p-1}[x_0(t), x_1(t), \dots, x_{n-1}(t), y_0(t), y_1(t), \dots, y_{p-1}(t)]$$

which can be written:

$$\mathbf{Y}(t+\tau) = F[\mathbf{X}(t), \mathbf{Y}(t)],$$

by putting $\mathbf{X} = (x_0, x_1, \dots, x_{n-1})$ and $\mathbf{Y} = (y_0, y_1, \dots, y_{p-1})$, called the *input* and *output vectors* respectively.

τ represents the time interval which separates the change in one of the input variables and the corresponding change in one of the output variables. In the case where τ corresponds to the clock period, the system is said to be *synchronous*.

4.2 STATEMENT OF EXERCISES AND PROBLEMS

4.E.1 Signal generator for the 6800 microprocessor

In order to function, the 6800 microprocessor requires three signals called Φ_1, Φ_2, and DBE. Φ_1 and Φ_2 are the two phases of the clock, DBE (data bus enable) signals that the data bus is active. A fourth signal, called bus Φ_2, is associated with these three; it activates circuits peripheral to the microprocessor (Figure 4.2).

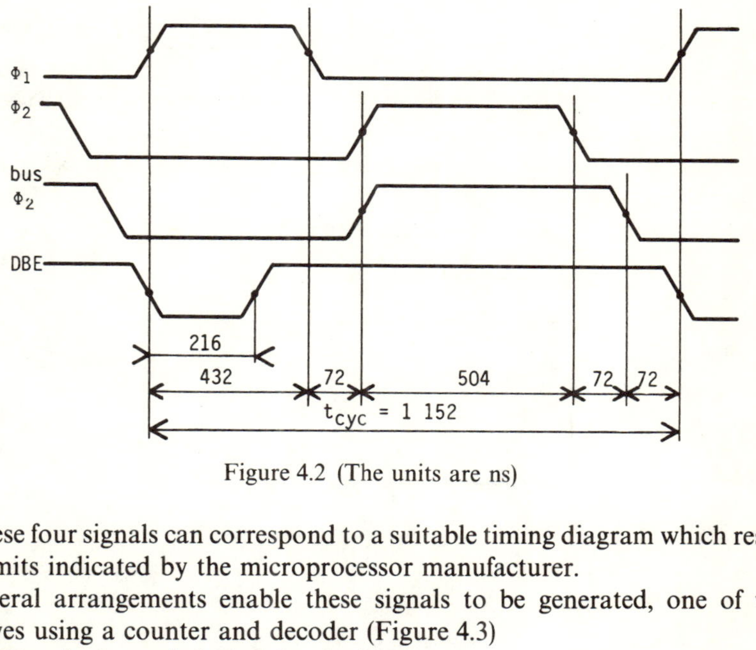

Figure 4.2 (The units are ns)

These four signals can correspond to a suitable timing diagram which respects the limits indicated by the microprocessor manufacturer.

Several arrangements enable these signals to be generated, one of which involves using a counter and decoder (Figure 4.3)

1. What is the period T of the clock?
2. How many states must be defined by the counter?
3. How many variables X_i are required by the counter?
4. Realize the four functions Φ_1, Φ_2, DBE, and bus Φ_2.
5. Give a diagram of the generator, attempting to avoid the different possible hazards.

4.E.2 Universal register

This is an array of n interconnected synchronous flip-flops into which data for n binary variables can be loaded and shifted to right or left. The universal register is represented by the symbol of Figure 4.4.

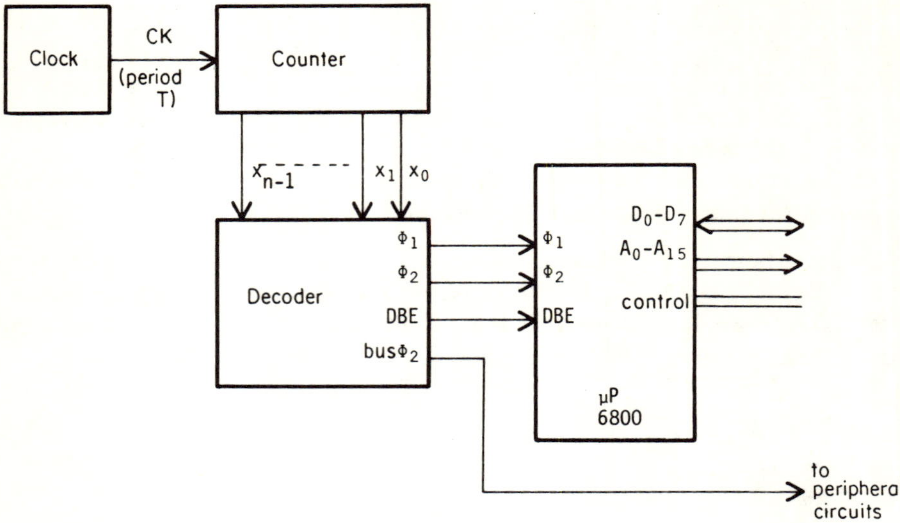

Figure 4.3

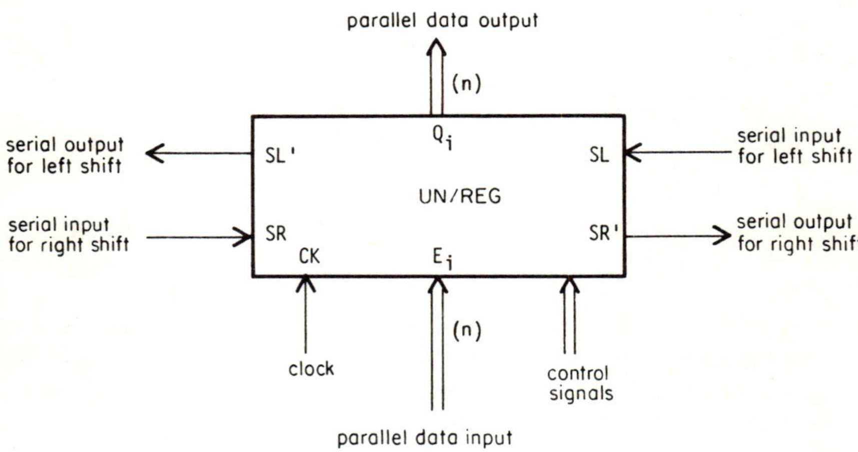

Figure 4.4

D	Q_n	Q_{n+1}
0	0	0
0	1	0
1	0	1
1	1	1

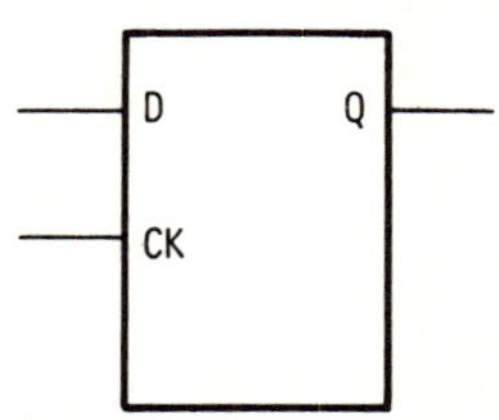

Figure 4.5

We shall realize the system using synchronous D flip-flops. Recall that a synchronous D flip-flop reproduces the state of the input D at the output Q, when the active edge of the clock appears. The operation is described by Figure 4.5, where Q_n and D are the states of the output and input before the arrival of the edge and Q_{n+1} the state of the output after.

On the other hand, from a dynamic point of view, the D flip-flop behaves as shown in the timing diagram of Figure 4.6.

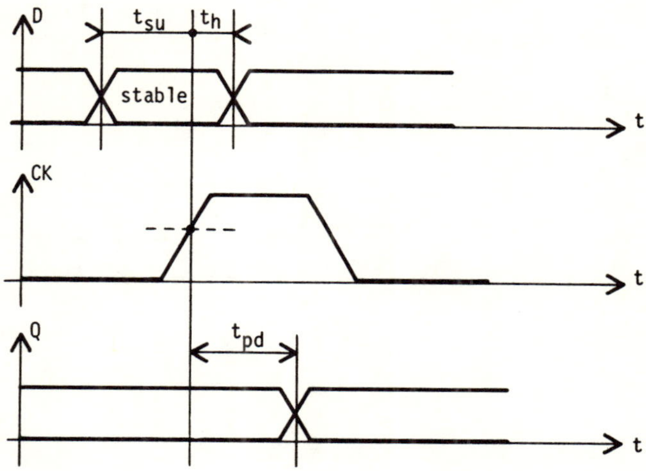

Figure 4.6

In Figure 4.6, t_{pd} is the propagation time (time taken for an input change to reach the output),

t_{su} is the set-up time,

t_h is the hold time.

This shows that D must be fixed during the time t_{su} before the active edge of the

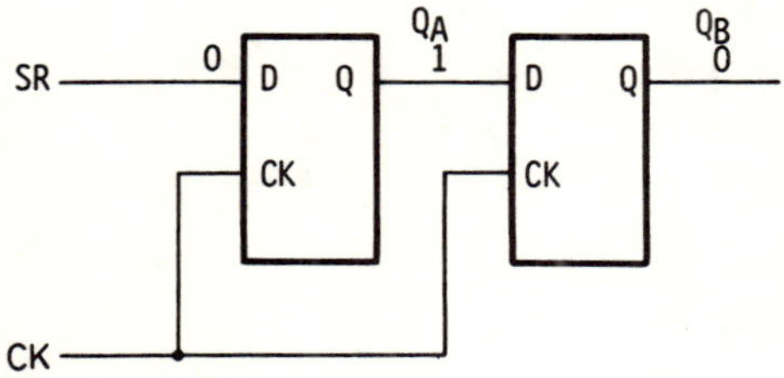

Figure 4.7

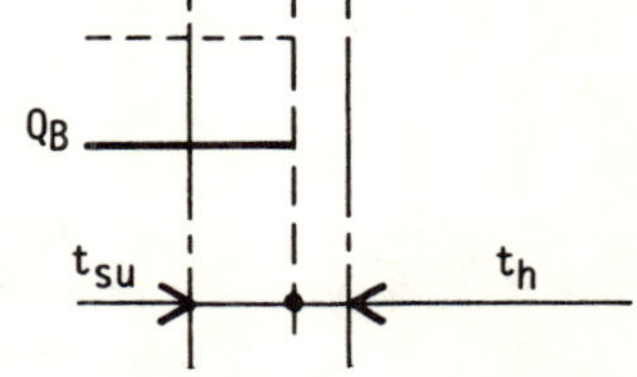

Figure 4.8

clock and be maintained during the time t_h after this edge. If these conditions are observed, the output Q correctly follows the input.

1. Two D flip-flops are connected and initialized as shown in Figure 4.7. What condition must be satisfied by t_{pd}, t_{su}, and t_h for the triplet SR, Q_A, $Q_B = 010$ to become 001? Justify by completing the timing diagram of Figure 4.8.
2. Complete the timing diagrams of Figure 4.9.

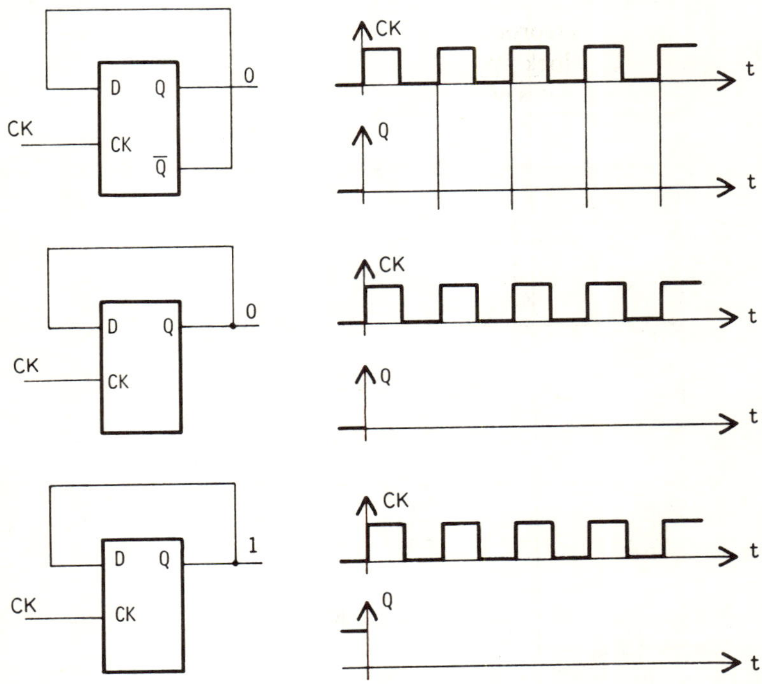

Figure 4.9

3. Consider the arrangement of Figure 4.10. What function does it realize?

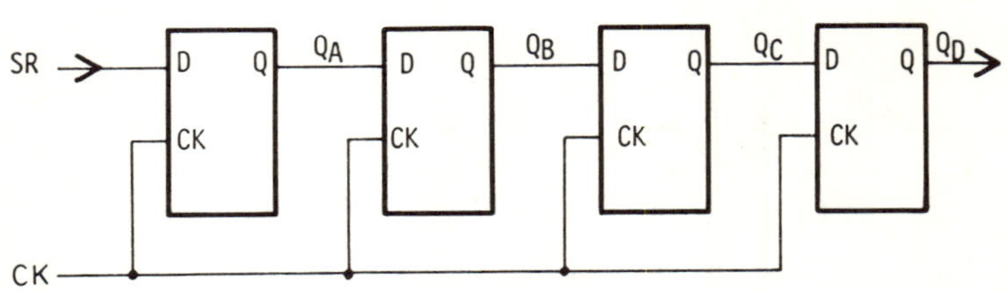

Figure 4.10

4. Consider the arrangement of Figure 4.11. What function does it realize?
5. The two previous functions are combined according to the diagram of Figure 4.12. What operation does M perform? Complete the diagram. Realize the operation M using NAND gates.

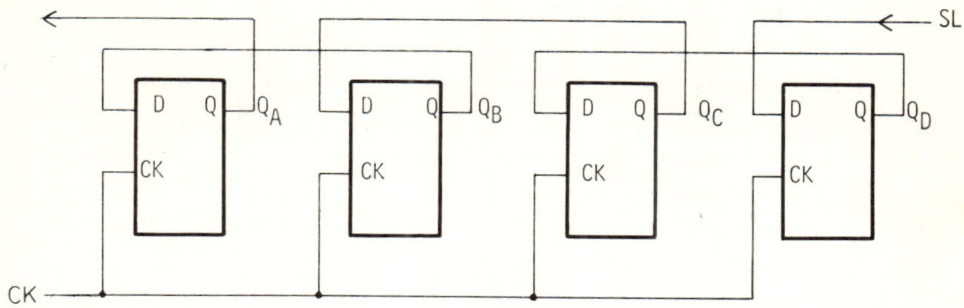

Figure 4.11

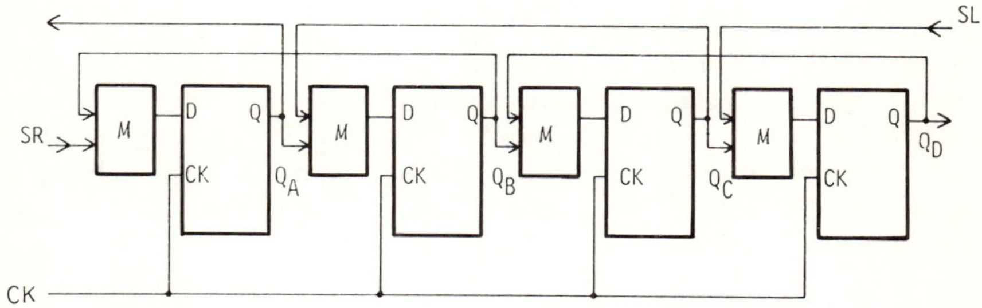

Figure 4.12

6. Is the condition of problem 1 always valid? Justify by completing the timing diagram of Figure 4.13.

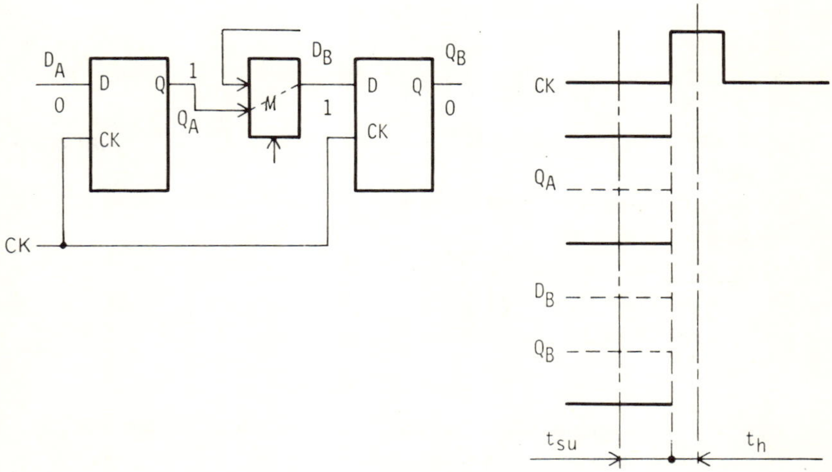

Figure 4.13

7. A word A, B, C, D may be loaded into Q_A, Q_B, Q_C, Q_D by adding a function to the previous arrangement, according to the diagram of Figure 4.14. What operation does M′ perform? Complete the diagram.

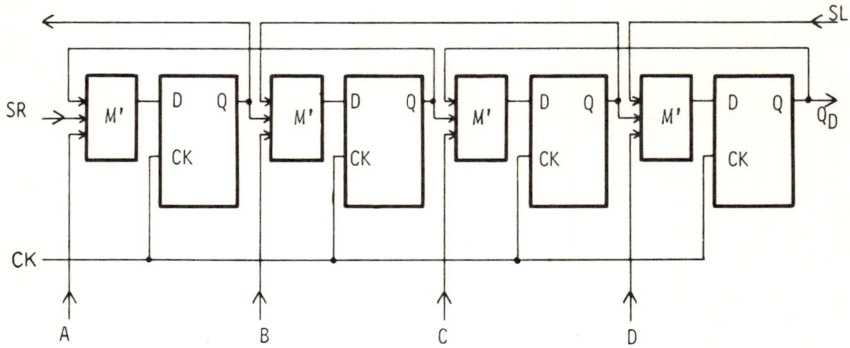

Figure 4.14

Realize the operation M′ using NAND gates.

8. In order to make the previous arrangement universal, an inhibition function is added: the contents of the flip-flops do not change in spite of the presence of a clock (dynamic halt). Give the final detailed arrangement by completing the diagram of Figure 4.15. The diagram obtained is that of a universal 4-bit register type 74S194 and 14194B.

4.E.3 Binary rate multiplier (BRM)

We are going to study a relatively little known component, which can be very useful in *serial* processing of data.

1. Consider a 4-bit binary counter: *d*, *c*, *b*, *a*. Complete the table of Figure 4.16. What do you notice?

2. The previous expressions are obtained as shown in Figure 4.17. Complete the timing diagrams of Figure 4.18.

3. The arrangement is modified as shown in Figure 4.19. Complete the timing diagrams of Figure 4.20

4. The inputs *D*, *C*, *B*, *A* are added and $S = D' + C' + B' + A'$ is connected as shown in Figure 4.21.

4a. Complete the timing diagrams of Figure 4.22.

4b. What do you notice?

4c. What name can be given to *T*?

4d. What weights can be assigned to *A*, *B*, *C*, *D*?

4e. In this case, what is the value of the expression:

$$N = p_3D + p_2C + p_1B + p_0A$$

where the p_i are the chosen weights and the + sign indicates the arithmetic sum?

4f. What kind of conversion has been realized?

5. Consider Figure 4.23 during *T* seconds.

The *pulse ratio r* is defined as the number of pulses present during *T* seconds compared with the number of unit intervals in *T* seconds.

5a. Establish the relation $r_{out} = f(r_{in}, N)$ by generalizing to a BRM of *n* bits.

5b. When can the value of *N* be changed?

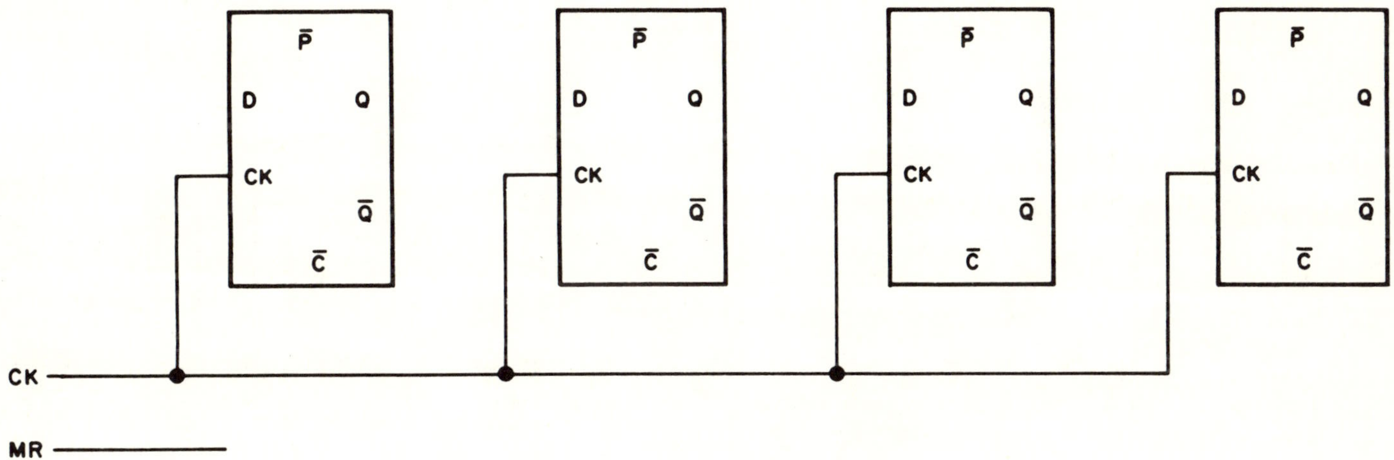

Figure 4.15

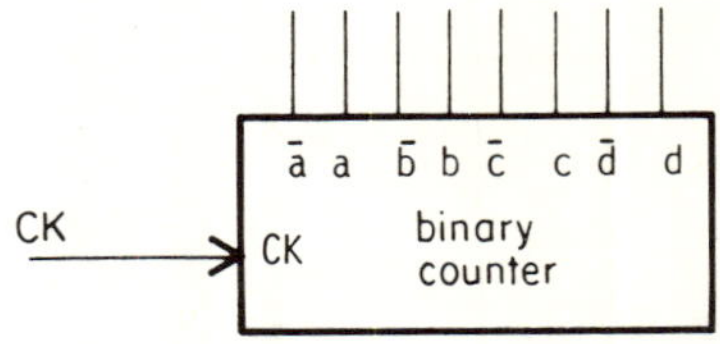

N	d (8)	c (4)	b (2)	a (1)	$\bar{a}$	$\bar{b}\cdot a$	$\bar{c}\cdot b\cdot a$	$\bar{d}\cdot c\cdot b\cdot a$
0	0	0	0	0				
1	0	0	0	1				
2	0	0	1	0				
3	0	0	1	1				
4	0	1	0	0				
5	0	1	0	1				
6	0	1	1	0				
7	0	1	1	1				
8	1	0	0	0				
9	1	0	0	1				
10	1	0	1	0				
11	1	0	1	1				
12	1	1	0	0				
13	1	1	0	1				
14	1	1	1	0				
15	1	1	1	1				

Figure 4.16

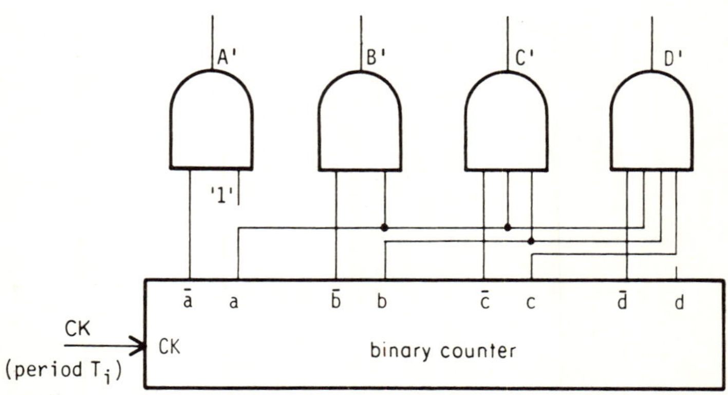

Figure 4.17

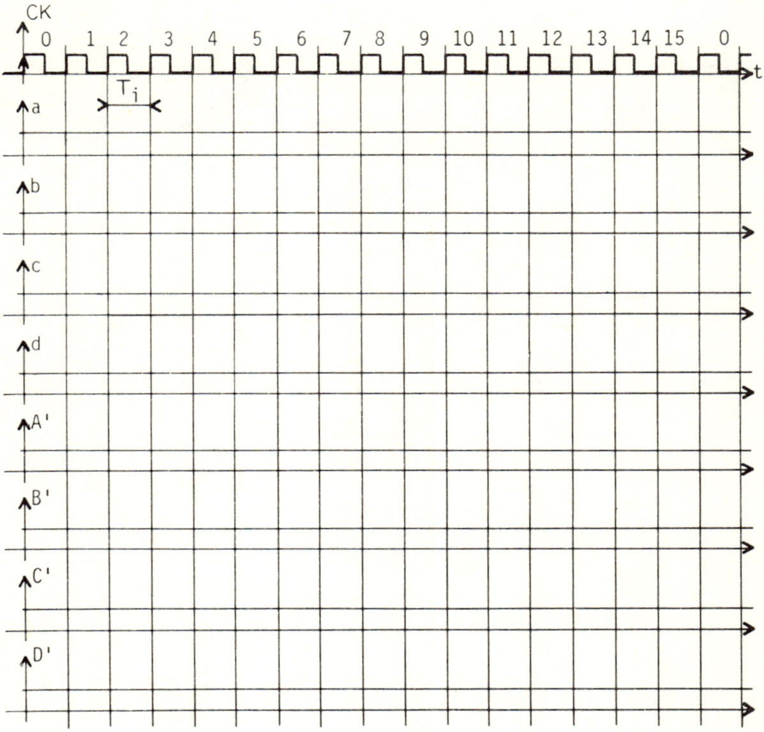

Figure 4.18

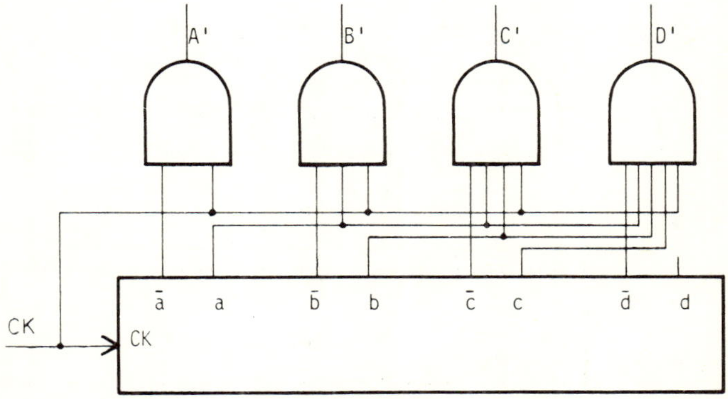

Figure 4.19

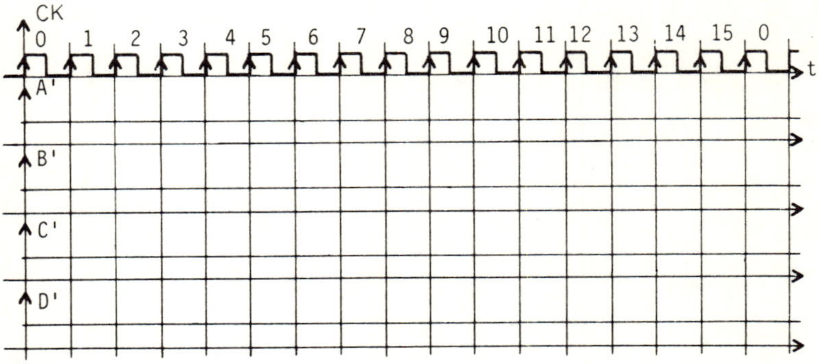

Figure 4.20

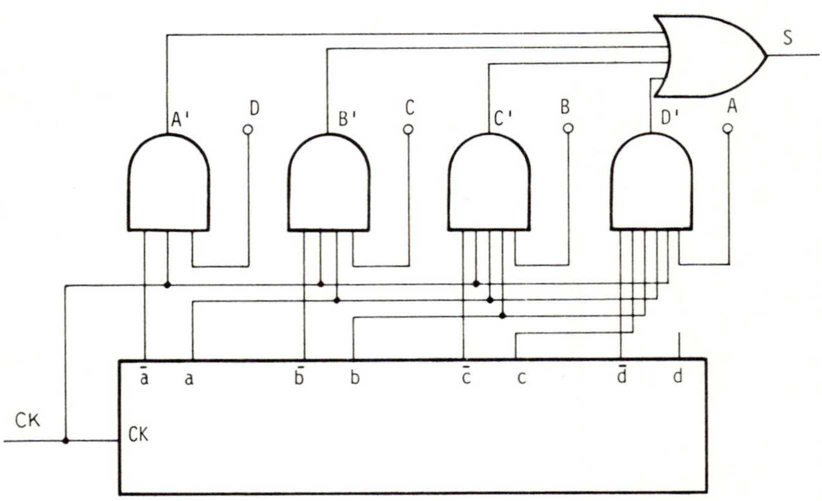

Figure 4.21

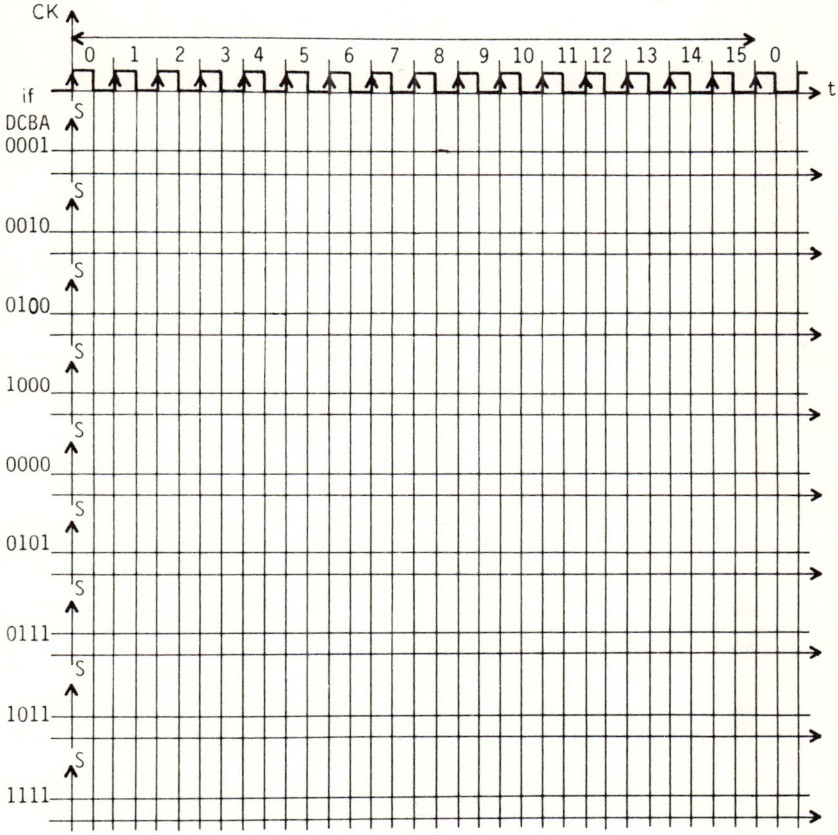

Figure 4.22

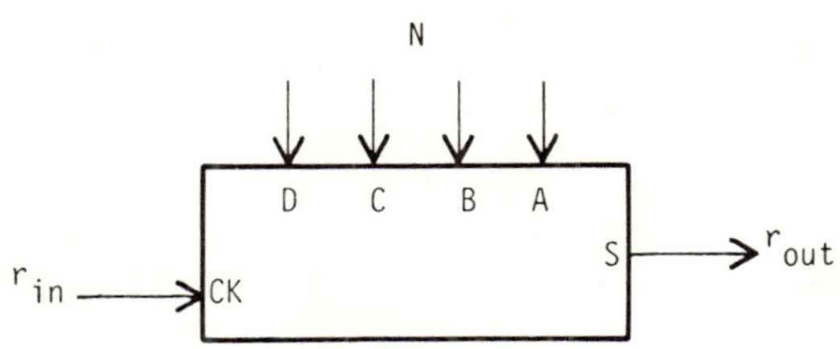

Figure 4.23

Application 1

6. Consider the diagram of Figure 4.24. Express $r_2 = g(r_{in}, X, Y,)$. What arithmetic operation has been performed?

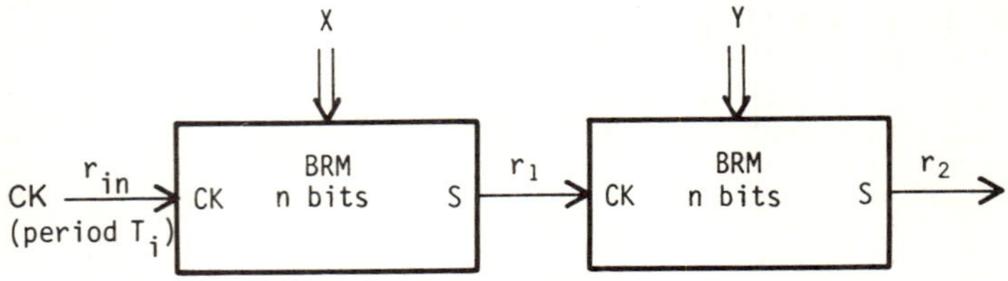

Figure 4.24

7. In order to obtain the result of this operation in numerical form, the arrangement of Figure 4.25 is realized.

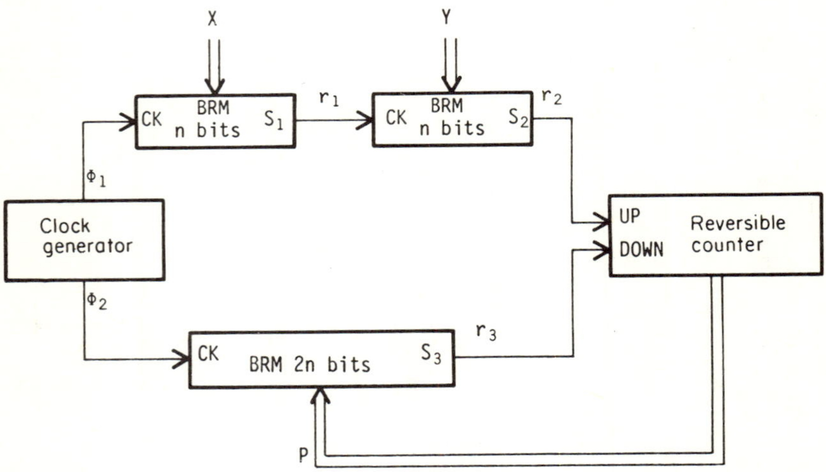

Figure 4.25

If P is the content of the reversible counter, the UP input allows a change to $P+1$ and the DOWN input to $P-1$. Φ_1 and Φ_2 are two phases of the clock, defined by Figure 4.26.

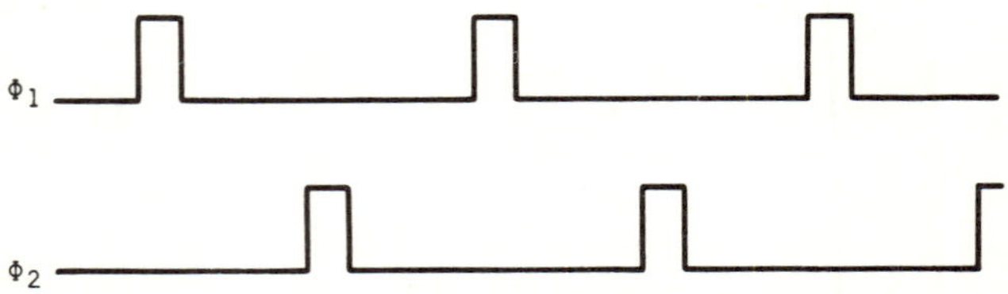

Figure 4.26

Explain the operation of this arrangement and write the relation between P and X and Y.

8. Show that this arrangement has a flaw which does not allow P to be obtained exactly.

In order to overcome this flaw, the arrangement of Figure 4.27 is interposed between the BRMs and the counter.

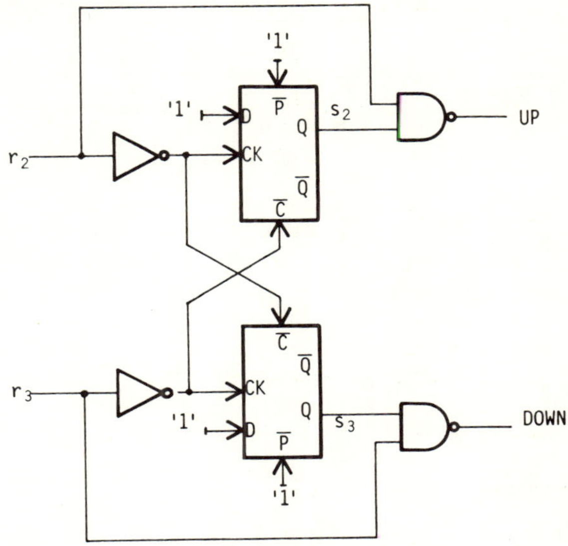

Figure 4.27

Explain its operation with the help of the timing diagram of Figure 4.28.

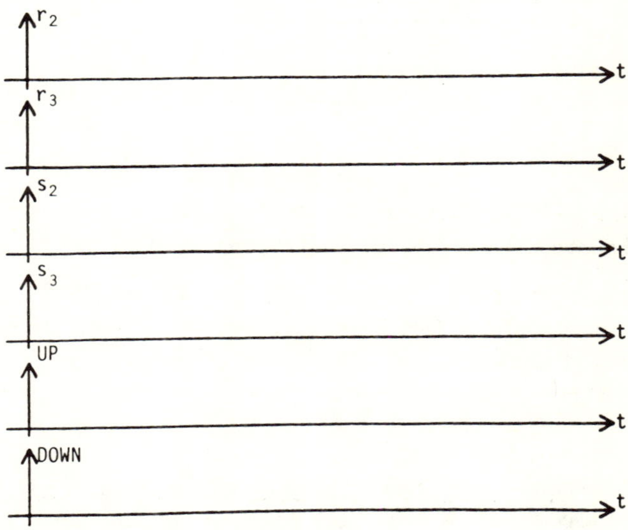

Figure 4.28

Application 2

9. Now consider the diagram of Figure 4.29.

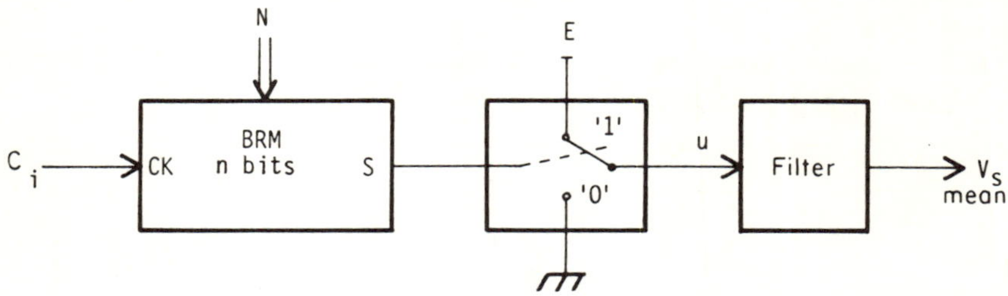

Figure 4.29

The clock C_i is as shown in Figure 4.30. The output S of the BRM controls an electronic switch such that when $S = 1$, $u = E$ volts and when $S = 0$, $u = 0$ volt. The purpose of the filter is to extract the mean component of the output signal u.

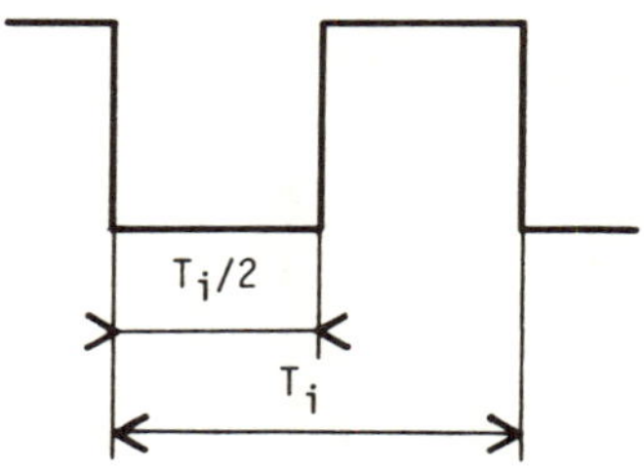

Figure 4.30

9a. Calculate V_{Smean} over the pseudo-period T.
9b. What function has been realized?
9c. When can the value of N be changed?

4.3 SOLUTIONS

4.S.1 Signal generator for the 6800 microprocessor

1. The clock period T is the HCF of 72, 216, 432, and 504, that is 72ns.
2. Knowing that $216 = 72 \times 3$, $432 = 72 \times 6$, $504 = 72 \times 7$, we notice that the period t_{cyc} of the 4 signals includes 16 of the clock periods T; so a counter with 16 states is appropriate.
3. Since $16 = 2^4$, we choose a 4-bit counter.
Let the four variables be x_3, x_2, x_1, x_0.
4. We could choose an integrated binary counter. However, it is interesting to choose a cyclic code which will avoid eventual decoding glitches. We shall select the well-known reflected binary code.

Figure 4.31 represents the timing diagrams which must be obtained.

Figure 4.31

Using four D flip-flops, the K maps of Figure 4.32 give:

$$D_3 = x_3\cdot x_1 + x_3\cdot x_0 + x_2\cdot\bar{x}_1\cdot\bar{x}_0,$$
$$D_2 = x_2\cdot\bar{x}_1 + x_2\cdot x_0 + \bar{x}_3\cdot x_1\cdot\bar{x}_0,$$
$$D_1 = x_1\cdot\bar{x}_0 + x_0\cdot(\overline{x_3\oplus x_2}),$$
$$D_0 = \overline{x_3\oplus x_2\oplus x_1},$$

from which:

$$D_3 = x_3\cdot\overline{\bar{x}_1\cdot\bar{x}_0} + x_2\cdot(\bar{x}_1\cdot\bar{x}_0),$$
$$D_2 = x_2\cdot\overline{x_1\cdot\bar{x}_0} + \bar{x}_3\cdot(x_1\cdot\bar{x}_0),$$
$$D_1 = \overline{\overline{x_1\cdot\bar{x}_0}\cdot\overline{x_0\cdot(x_3\oplus\bar{x}_2)}},$$
$$D_0 = (x_3\oplus\bar{x}_2)\cdot x_1$$

The diagram of the counter is given in Figure 4.33.

The operation of M in Figure 4.33 is that of a 2-input multiplexer, used as a function generator (see Chapter 1). In particular, two multiplexers allow the generation of $D_0 = x_3\oplus x_2\oplus x_1$.

The interest of this solution is that it requires only two 7474s, one 74157,

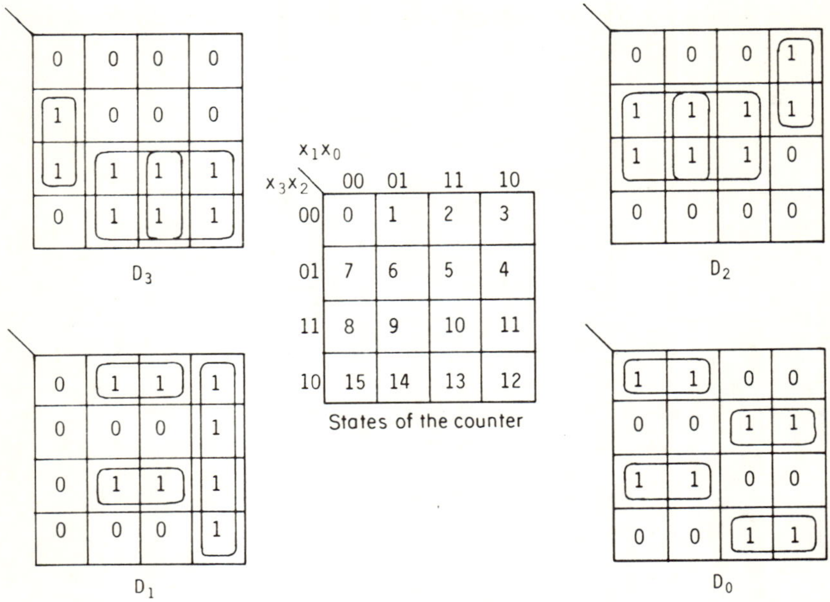

Figure 4.32

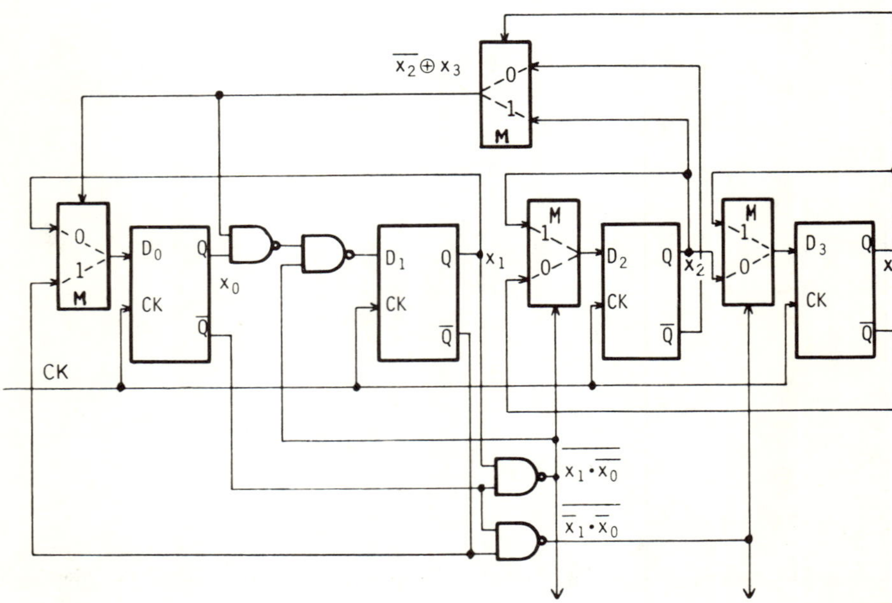

Figure 4.33

and one 7400, that is four *D* flip-flops, four 2-input multiplexers, and four 2-input NAND gates.

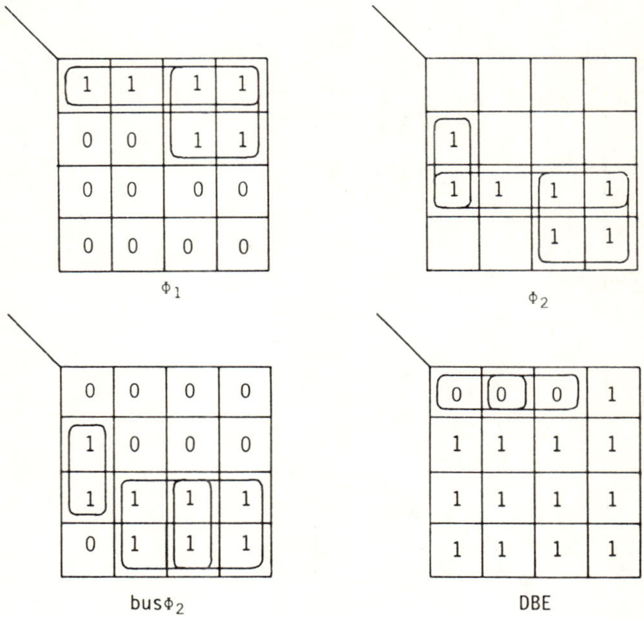

Figure 4.34

Four K maps (Figure 4.34) give:

$$\Phi_1 = \bar{x}_3 \cdot \bar{x}_2 + \bar{x}_3 \cdot x_1,$$
$$\text{bus } \Phi_2 = x_3 \cdot x_2 + x_3 \cdot x_1 + x_2 \cdot \bar{x}_1 \cdot \bar{x}_0,$$
$$\Phi_2 = x_3 \cdot x_1 + x_3 \cdot x_0 + x_2 \cdot \bar{x}_1 \cdot \bar{x}_0,$$
$$\text{DBE} = (x_3 + x_2 + x_1) \cdot (x_3 + x_2 + \bar{x}_0),$$

from which:

$$\Phi_1 = \overline{x_3 + \overline{x_2 \cdot \bar{x}_1}},$$
$$\Phi_2 = \overline{\bar{x}_3 + \overline{x_2 + x_1}} + \overline{x_2 + \overline{\bar{x}_1 \cdot \bar{x}_0}},$$
$$\text{bus } \Phi_2 = D_3,$$
$$\text{DBE} = \overline{x_3 + x_2 \cdot x_1 \cdot \bar{x}_0}.$$

5. The diagram of the decoder is given in Figure 4.35.

4.S.2 Universal register

1. The timing diagram of Figure 4.36 shows that the condition to be observed is:

$$t_h < t_{pd}.$$

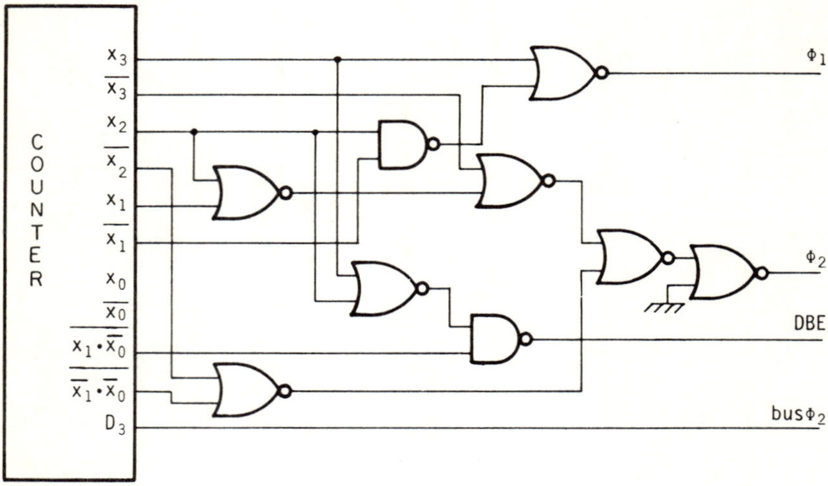

Figure 4.35

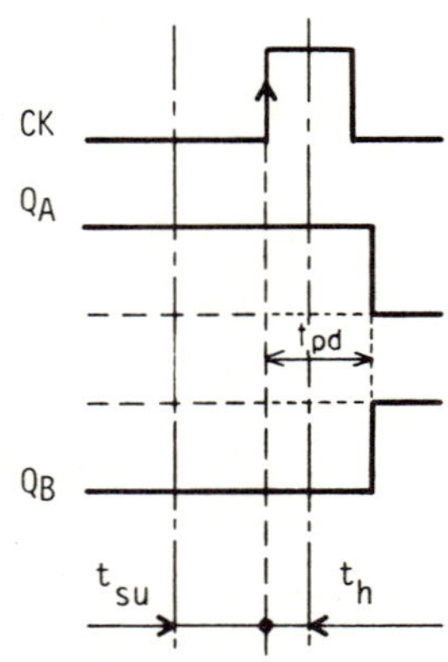

Figure 4.36

2. The timing diagrams of Figure 4.37 show that the *D* flip-flop can operate as a divide-by-two or be inhibited to '0' or '1'.

3. The function realized is a shift to the right of the data contained in the array of four flip-flops, called a *register shift right*.

4. The function realized is a shift to the left of the data.

5. The operation of M is that of a 2-input multiplexer.

Figure 4.38 gives the complete diagram, in which:

SR is the right serial input
SL is the left serial input
S_0 is the control which, acting on the multiplexer, allows the direction of shifting to be chosen.

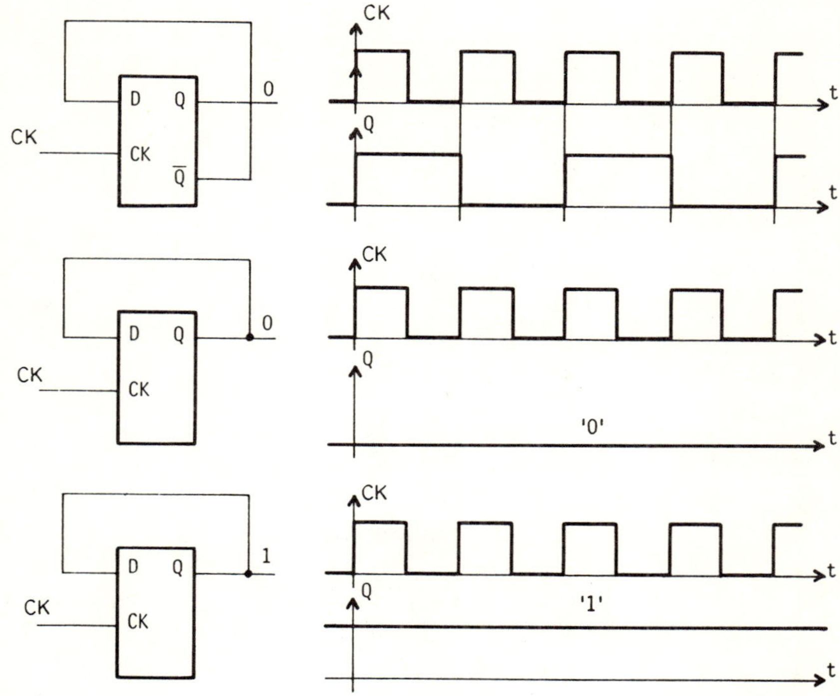

Figure 4.37

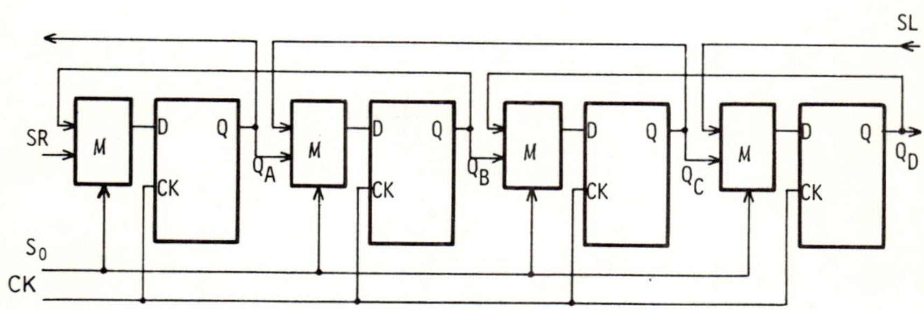

Figure 4.38

The operation of M is defined by Figure 4.39 (see chapter 1).

The table of operations is given in Figure 4.40.

6. The timing diagram of Figure 4.41 shows that the condition of problem 1 is even better observed since:

$$t_h < t_{pd} + \tau,$$

where τ is the propagation delay of the muliplexer.

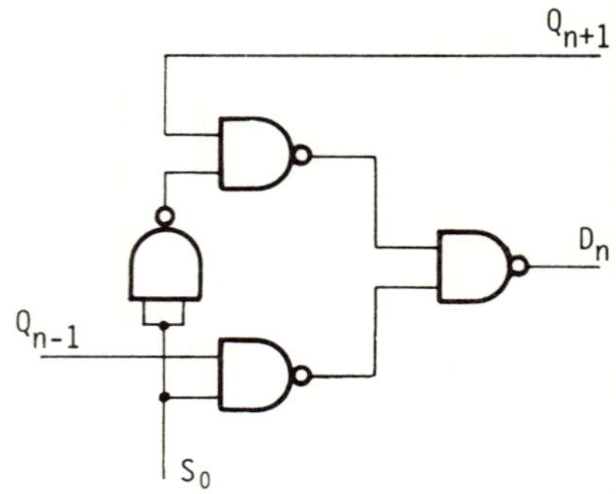

Figure 4.39

S_0	function realized
0	shifting $Q_A \leftarrow Q_D$ called shift left
1	shifting $Q_A \rightarrow Q_D$ called shift right

Figure 4.40

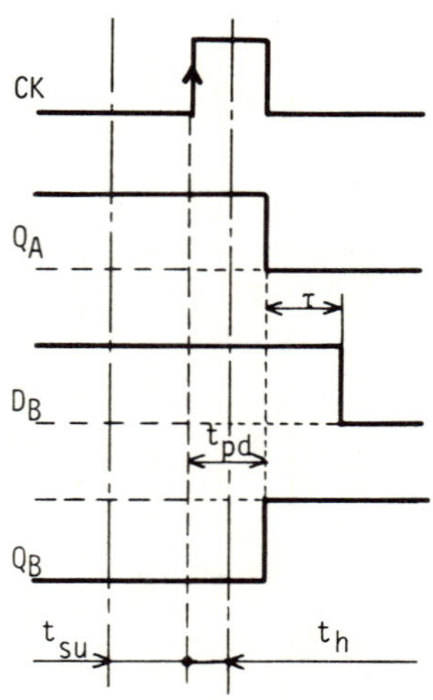

Figure 4.41

7. The operation of M′ is that of 4-input multiplexer of which one input is not used.

Figure 4.42 gives the complete diagram, in which S_1 and S_0 are the controls which, acting on the multiplexer, allow the direction of shifting to be chosen.

Docklands Library & Learning Centre
Issued items receipt

Customer name: INACIO DA COSTA, EDILSON

Title: Practical digital electronics /
ID: 24961404
Due: 15/12/2010 23:59
Circulation system messages:
Loan performed

Total items this session: 1
17/11/2010 17:31
Total number of loans: 3
Overdue: 0

Please keep this receipt.

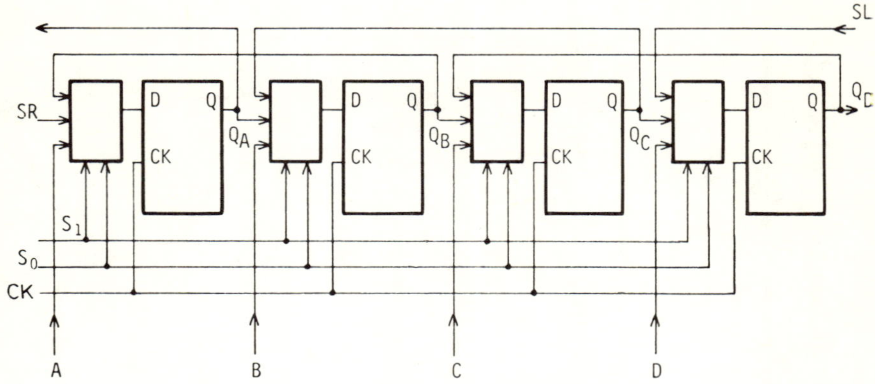

Figure 4.42

The operation of M′ is defined by Figure 4.43.

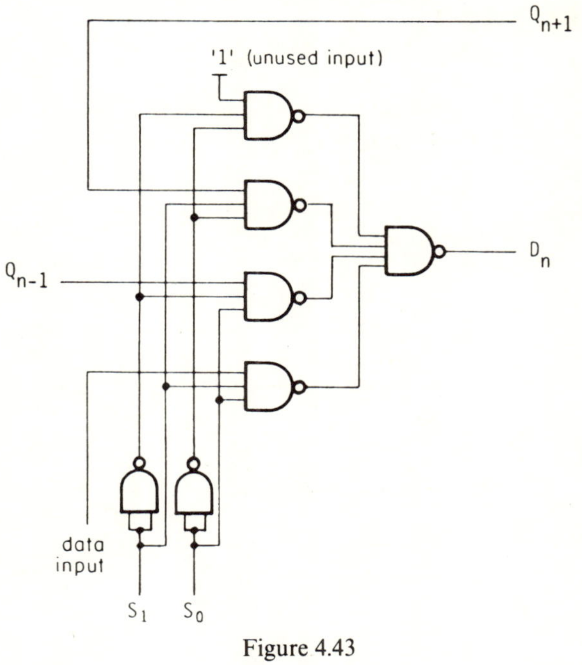

Figure 4.43

The table of operations is given in Figure 4.44.

S_1	S_0	function realized
0	0	none
0	1	shifting $Q_A \rightarrow Q_D$
1	0	shifting $Q_A \leftarrow Q_D$
1	1	loading $A, B, C, D \rightarrow Q_A, Q_B, Q_C, Q_D$

Figure 4.44

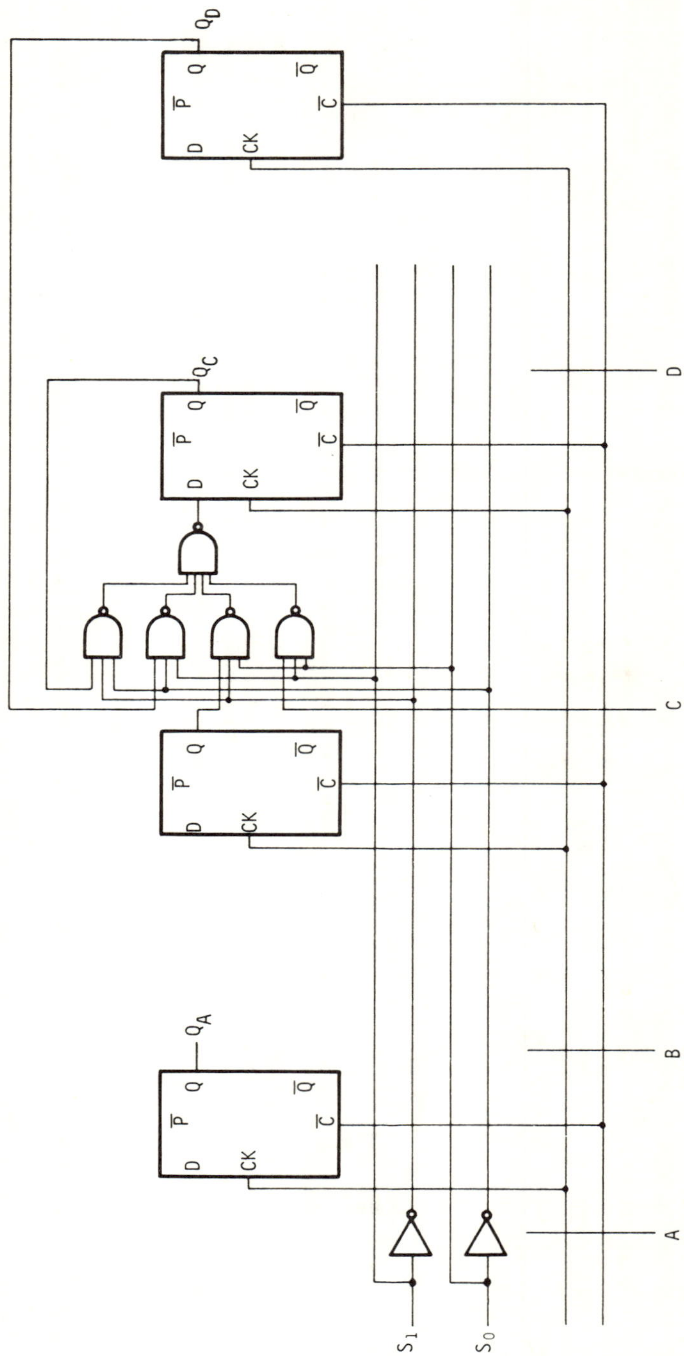

Figure 4.45

8. The final diagram is given in Figure 4.45. In order to obtain a clear diagram, only one stage is represented: it uses the multiplexer of Figure 4.43 where the unused input is connected to the output Q_n of the flip-flop of level *n*. In this case, the table of operations is given in Figure 4.46.

S_1	S_0	function realized
0	0	inhibition
0	1	shifting $Q_A \rightarrow Q_D$
1	0	shifting $Q_A \leftarrow Q_D$
1	1	loading $A, B, C, D \rightarrow Q_A, Q_B, Q_C, Q_D$

Figure 4.46

4.S.3 Binary rate multiplier (BRM)

1. The completed table is given in Figure 4.47. Only the '1' values of the functions are represented, the other combinations correspond to '0'.

Counting the '1's, we notice that there are 8 in the first column, 4 in the second, 2 in the third, and 1 in the fourth, and that the '1's are regularly distributed. Also, a '1' for each of the functions created corresponds to each combination of the

N	d (8)	c (4)	b (2)	a (1)	$\bar{a}$	$\bar{b} \cdot a$	$\bar{c} \cdot b \cdot a$	$\bar{d} \cdot c \cdot b \cdot a$
0	0	0	0	0	1			
1	0	0	0	1		1		
2	0	0	1	0	1			
3	0	0	1	1			1	
4	0	1	0	0	1			
5	0	1	0	1		1		
6	0	1	1	0	1			
7	0	1	1	1				1
8	1	0	0	0	1			
9	1	0	0	1		1		
10	1	0	1	0	1			
11	1	0	1	1			1	
12	1	1	0	0	1			
13	1	1	0	1		1		
14	1	1	1	0	1			
15	1	1	1	1				

Figure 4.47

variables d, c, b, a, except for combination 15 where the four functions have the '0' state.

2. The completed timing diagrams are given in Figure 4.48. The distribution of pulses indicated previously can be clearly seen.

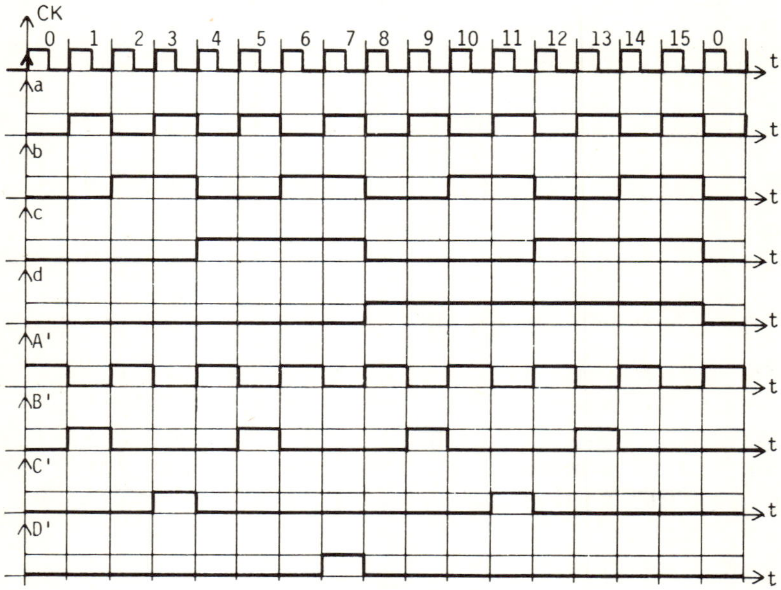

Figure 4.48

3. The completed timing diagrams are given in Figure 4.49. They are the previous diagrams with the output pulses timed by the clock.

4a. The completed timing diagrams are given in Figure 4.50.

4b. We notice that the number of pulses present at the output S, during T seconds, depends on the energized input. Thus:

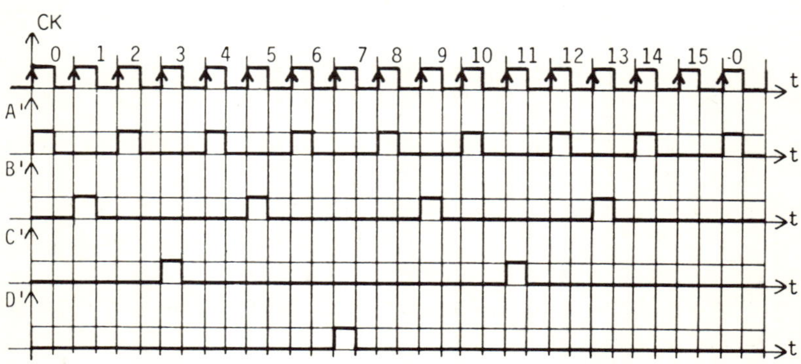

Figure 4.49

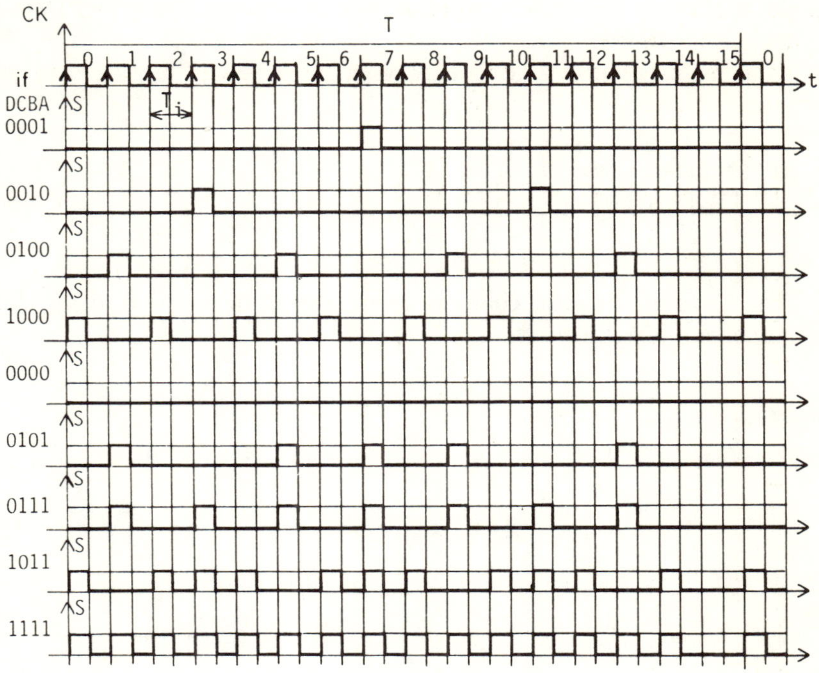

Figure 4.50

when $A = 1$ there is 1 pulse on S,
when $B = 1$ there are 2 pulses on S,
when $C = 1$ there are 4 pulses on S,
when $D = 1$ there are 8 pulses on S.

Also, it is possible to energize several inputs simultaneously; in this case, S produces the *sum of the pulses* caused by each energized input. Hence, for the input combinations DCBA:

0101 corresponds to 5 pulses on S,
0111 corresponds to 7 pulses on S,
1011 corresponds to 11 pulses on S,
1111 corresponds to 15 pulses on S,

during T seconds, that is when 16 pulses are *regularly* presented at the clock input.

4c. T can be called the *pseudo-period*; in the general case, $T = 2^n T_i$.

4d. It is evident that weights of 1-2-4-8 should be taken by inputs A, B, C, D.

4e. In this case, $N = 8D + 4C + 2B + A$.

4f. In this way a conversion from a numerical value to a number of pulses has been realized.

5a. In our case, T seconds represent 16 clock periods. Therefore during this time there are 16 pulses present on the CK input; it follows that $r_{in} = 16/16 = 1$.

At the output S, the number of pulses present is proportional to r_{in} and to N, as we have seen above; hence:

$$r_{out} = r_{in} \times \frac{N}{16}$$

In the general case, with an n-bit BRM, we obtain:

$$r_{out} = r_{in} \times \frac{N}{2^n}$$

5b. The value of N can be changed during the (2^n)th period of the cycle T, since it does not affect the output.

Application 1

6. We have:

$$r_{in} = 1,$$

$$r_1 = r_{in} \times \frac{X}{2^n} = 1 \times \frac{X}{2^n}$$

$$r_2 = r_1 \times \frac{Y}{2^n} = \frac{X \times Y}{2^{2n}}$$

This expression shows that the number of pulses present at the output of the second BRM is equal to the *arithmetic product* of the values of X and Y, during a period $T' = 2^n T$, that is the product $X \times Y$ is counted with respect to 2^{2n} pulses.

7. We have seen that $r_2 = \dfrac{X \times Y}{2^{2n}}$

Similarly:

$$r_3 = \frac{P}{2^{2n}}.$$

The counter adds $+1$ to its contents when a pulse appears on S_2, and -1 when a pulse appears on S_3. Now the two phases of the clock prevent pulses on the UP and DOWN inputs from coinciding; so Φ_1, by way of S_2, controls UP, and Φ_2, by way of S_3, controls DOWN.

After a transient region, during which P increases, an *equilibrium region* is established such that $r_2 = r_3$. In this case:

$$\frac{X \times Y}{2^{2n}} = \frac{P}{2^{2n}}, \quad \text{from which} \quad P = X \times Y.$$

So we obtain the numerical value of the product $X \times Y$, coded with $2n$ bits, in P.

8. In equilibrium, one of the following situations applies:

— one pulse on UP results in $P = X \times Y + 1$, the pulse on DOWN, which follows, results in $P = X \times Y$;

—one pulse on UP results in $P = X \times Y$, the pulse on DOWN, which follows, results in $P = X \times Y - 1$.

In other words, the value of P *oscillates* about the required product $X \times Y$.

The arrangement of Figure 4.27 has the purpose of transmitting to the UP (DOWN) output two or more consecutive pulses applied to the r_2 (r_3) input, without a pulse appearing on r_3 (r_2).

As soon as a pulse appears on r_3 (r_2), the UP (DOWN) output is deactivated and the DOWN (UP) output activated for the pulses successively applied to r_3(r_2). When the pulses appear alternately on r_2 and r_3, the UP and DOWN outputs are deactivated.

The operation appears on the timing diagrams of Figure 4.51.

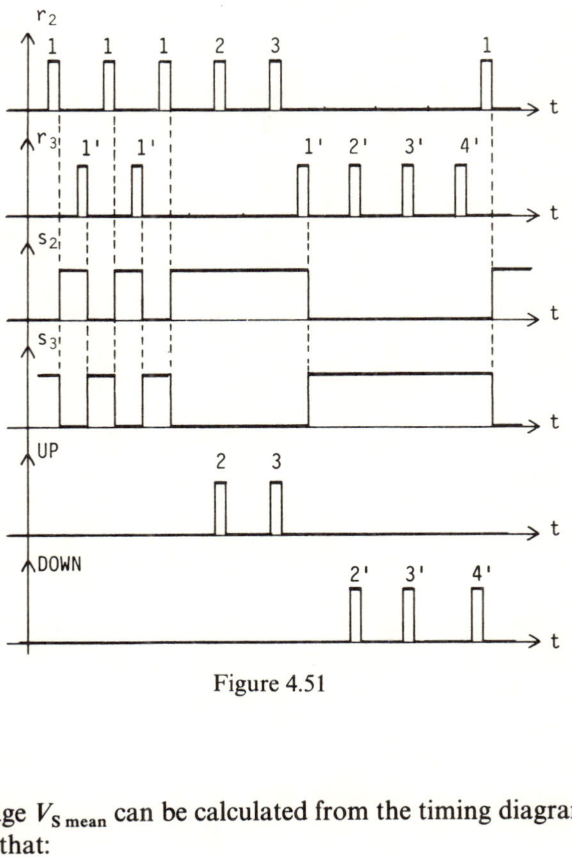

Figure 4.51

Application 2

9a. The voltage $V_{S\,\text{mean}}$ can be calculated from the timing diagrams of Figure 4.52. It is clear that:

$$V_{\text{Smean}} = \frac{E \times N \times T_i/2}{2^n \times T_i} = \frac{E \times N}{2^{n+1}}.$$

9b. We have realized a *serial digital to analog conversion.*

9c. The value of N can be changed during the (2^n)th period of cycle T, as we have seen in question 5b.

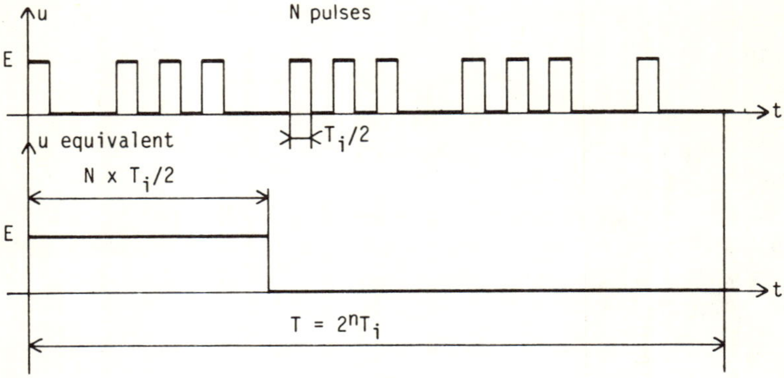

Figure 4.52 Example for $N = 11 = (8 + 2 + 1)$.

Figures 4.53, 4.54, and 4.55 show a commercial binary BRM and two decimal BRMs respectively.

STATE AND/OR RATE FUNCION TABLE (See Note A)

INPUTS											OUTPUTS			
			BINARY RATE								LOGIC LEVEL OR NUMBER OF PULSES			
CLEAR	ENABLE	STROBE	F	E	D	C	B	A	NUMBER OF CLOCK PULSES	UNITY/ CASCADE	Y	Z	ENABLE	NOTES
H	X	H	X	X	X	X	X	X	X	H	L	H	H	B
L	L	L	L	L	L	L	L	L	64	H	L	H	1	C
L	L	L	L	L	L	L	L	H	64	H	1	1	1	C
L	L	L	L	L	L	L	H	L	64	H	2	2	1	C
L	L	L	L	L	L	H	L	L	64	H	4	4	1	C
L	L	L	L	L	H	L	L	L	64	H	8	8	1	C
L	L	L	L	H	L	L	L	L	64	H	16	16	1	C
L	L	L	H	L	L	L	L	L	64	H	32	32	1	C
L	L	L	H	H	H	H	H	H	64	H	63	63	1	C
L	L	L	H	H	H	H	H	H	64	L	H	63	1	D
L	L	L	H	L	H	L	L	L	64	H	40	40	1	E

NOTES:

A. H = High level, L = Low level, X = irrelevant. All remaining entries are numeric counts.

B. This is a simplified illustration of the clear function. The states of clock and strobe can affect the logic level of Y and Z. A low unity/cascade will cause output Y to remain high.

C. Each rate illustrated assumes a constant value of rate inputs; however, these illustrations in no way prohibit variable rate inputs.

D. Unity/cascade is used to inhibit output Y.

E. $f_{out} = \frac{M - f_{in}}{64} = \frac{(8+32)f_{in}}{64} = \frac{40f_{in}}{64} = 0.625f_{in}$

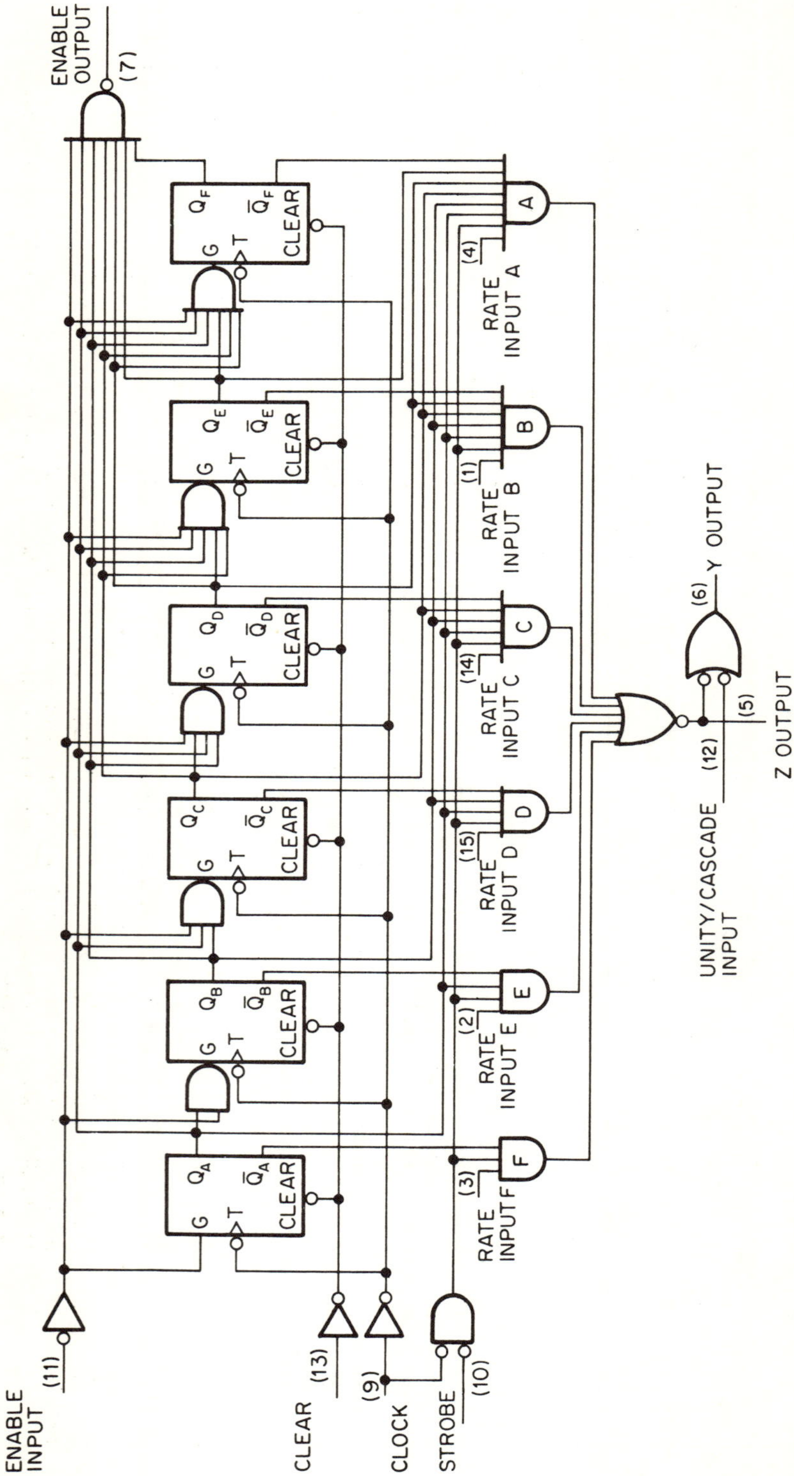

ENABLE OUTPUT
(7)
ENABLE INPUT
(11)
CLEAR
(13)
CLOCK
(9)
STROBE
(10)
RATE INPUT A
(4)
RATE INPUT B
(1)
RATE INPUT C
(14)
RATE INPUT D
(15)
RATE INPUT E
(2)
RATE INPUT F
(3)
A
B
C
D
E
F
Y OUTPUT
(6)
Z OUTPUT
(5)
UNITY/CASCADE INPUT
(12)

Figure 4.53

STATE AND/OR RATE FUNCION TABLE (See Note A)

INPUTS									OUTPUTS			
									LOGIC LEVEL OR NUMBER OF PULSES			
CLEAR	ENABLE	STROBE	BCD RATE D	C	B	A	NUMBER OF CLOCK PULSES	UNITY/ CASCADE	Y	Z	ENABLE	NOTES
H	X	H	X	X	X	X	X	H	L	H	H	B
L	L	L	L	L	L	L	10	H	L	H	1	C
L	L	L	L	L	L	H	10	H	1	1	1	C
L	L	L	L	L	H	L	10	H	2	2	1	C
L	L	L	L	L	H	H	10	H	3	3	1	C
L	L	L	L	H	L	L	10	H	4	4	1	C
L	L	L	L	H	L	H	10	H	5	5	1	C
L	L	L	L	H	H	L	10	H	6	6	1	C
L	L	L	L	H	H	H	10	H	7	7	1	C
L	L	L	H	L	L	L	10	H	8	8	1	C
L	L	L	H	L	L	H	10	H	9	9	1	C
L	L	L	H	L	H	L	10	H	8	8	1	C, D
L	L	L	H	L	H	H	10	H	9	9	1	C, D
L	L	L	H	H	L	L	10	H	8	8	1	C, D
L	L	L	H	H	L	H	10	H	9	9	1	C, D
L	L	L	H	H	H	L	10	H	8	8	1	C, D
L	L	L	H	H	H	H	10	H	9	9	1	C, D
L	L	L	H	L	L	H	10	L	H	9	1	E

NOTES:
A. H = High level, L = Low level, X = irrelevant. All remaining entries are numeric counts.
B. This is a simplified illustration of the clear function. The states of clock and strobe can affect the logic level of Y and Z. A low unity/cascade will cause output Y to remain high.
C. Each rate illustrated assumes a constant value of rate inputs; however, these illustrations in no way prohibit variable rate inputs.
D. These input conditions exceed the range of the decimal rate inputs.
E. Unity/cascade is used to inhibit output Y.

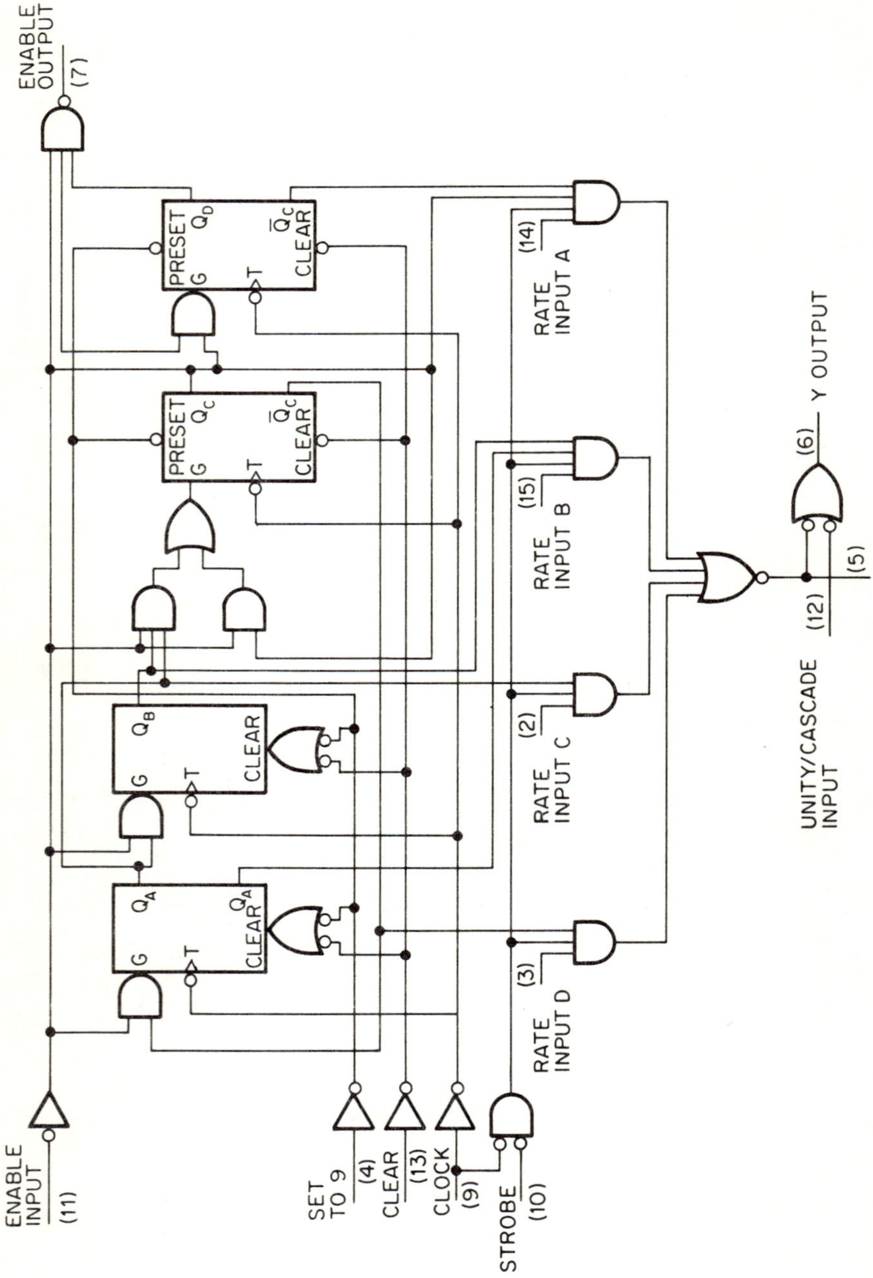
ENABLE OUTPUT
(7)
PRESET
G
T
CLEAR
RATE INPUT A
(14)
RATE INPUT B
(15)
RATE INPUT C
(2)
RATE INPUT D
(3)
Y OUTPUT
(6)
Z OUTPUT
(5)
(12)
UNITY/CASCADE INPUT
ENABLE INPUT
(11)
SET TO 9
(4)
CLEAR
(13)
CLOCK
(9)
STROBE
(10)

Figure 4.54

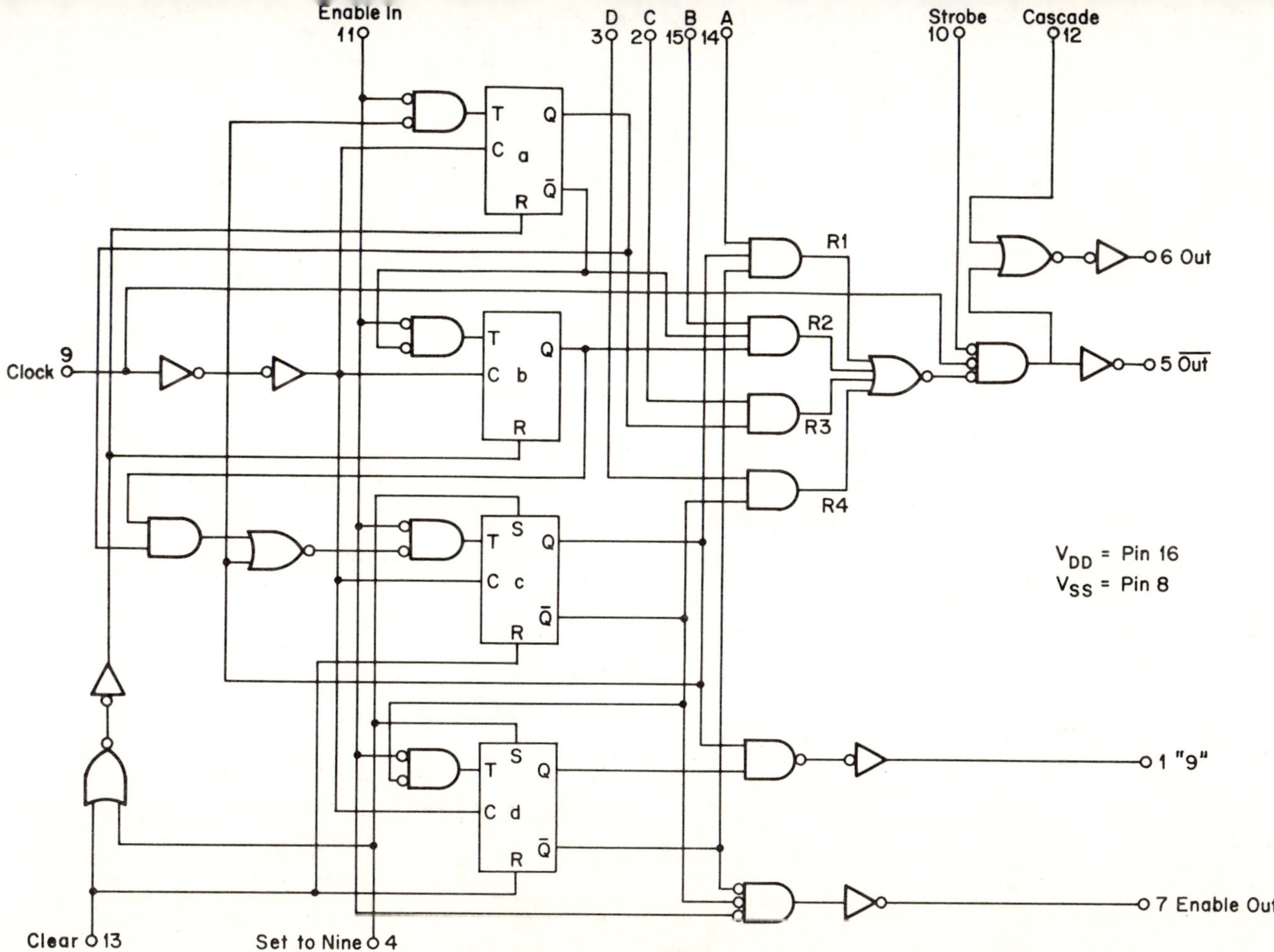

Enable In
11
D
3
C
2
B
15
A
14
Strobe
10
Cascade
12
T
C
a
Q
$\bar{Q}$
R
T
C
b
Q
R
S
T
C
c
Q
$\bar{Q}$
R
S
T
C
d
Q
$\bar{Q}$
R
R1
R2
R3
R4
6 Out
5 $\overline{\text{Out}}$
9
Clock
V_{DD} = Pin 16
V_{SS} = Pin 8
1 "9"
7 Enable Out
Clear 13
Set to Nine 4

Figure 4.55(a)

TRUTH TABLE

MC 14527B INPUTS										OUTPUT LOGIC LEVEL NUMBER OF PULSES			
D	C	B	A	No. of Clock Pulses	E_{in}	STROBE	CASCADE	CLEAR	SET	OUT	$\overline{OUT}$	E_{out}	"9"
0	0	0	0	10	0	0	0	0	0	0	1	1	1
0	0	0	1	10	0	0	0	0	0	1	1	1	1
0	0	1	0	10	0	0	0	0	0	2	2	1	1
0	0	1	1	10	0	0	0	0	0	3	3	1	1
0	1	0	0	10	0	0	0	0	0	4	4	1	1
0	1	0	1	10	0	0	0	0	5	5	5	1	1
0	1	1	0	10	0	0	0	0	0	6	6	1	1
0	1	1	1	10	0	0	0	0	0	7	7	1	1
1	0	0	0	10	0	0	0	0	0	8	8	1	1
1	0	0	1	10	0	0	0	0	0	9	9	1	1
1	0	1	0	10	0	0	0	0	0	8	8	1	1
1	0	1	1	10	0	0	0	0	0	9	9	1	1
1	1	0	0	10	0	0	0	0	0	8	8	1	1
1	1	0	1	10	0	0	0	0	0	9	9	1	1
1	1	1	0	10	0	0	0	0	0	8	8	1	1
1	1	1	1	10	0	0	0	0	0	9	9	1	1
×	×	×	×	10	1	0	0	0	0	—	—	—	—
×	×	×	×	10	0	1	0	0	0	0	1	1	1
×	×	×	×	10	0	0	1	0	0	1	0	1	1
1	×	×	×	10	0	0	0	1	0	10	10	1	0
0	×	×	×	10	0	0	0	1	0	0	1	1	0
×	×	×	×	10	0	0	0	0	1	0	1	0	1

$x =$ Don't care

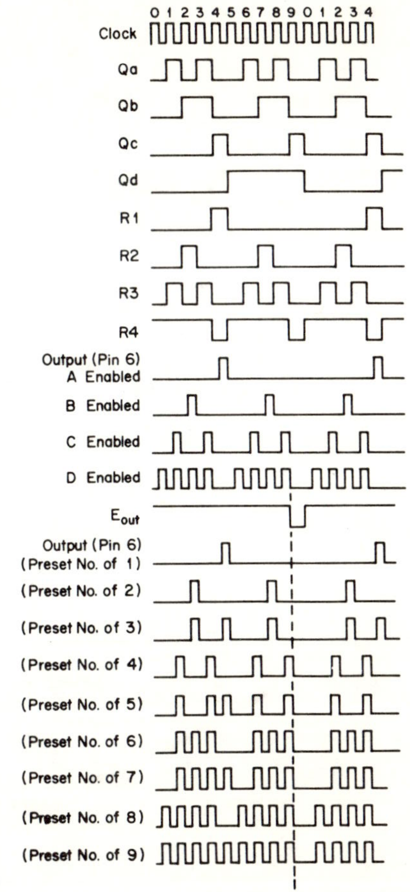

Figure 4.55(b)

5 Arithmetic Systems

5.1 SUMMARY OF THEORY

Bases of numbers

With the commonly used positional notation, every number X can be written in the form:

$$X = x_{n-1}x_{n-2}\ldots x_1 x_0 \cdot x_{-1}\ldots x_{-p}$$

where x_i are the figures 0, 1,...$(x_{n-1}\ldots x_0)$ is called the *integer part*; $(x_{-1}\ldots x_{-p})$ is called the *fractional part*.

The index i is called the *rank* of the digit.

The number *base* B involves B symbols: the figures 0, 1, 2,..., $(B-1)$. It should be noted that above 9, the letters A, B,... are used. All x_i of the number X must be symbols belonging to the base.

Since base ten is commonly used, the number $X_{(B)}$, in base b, can be written in base ten:

$$X_{(\text{ten})} = x_{n-1}B^{n-1} + x_{n-2}B^{n-2} + \ldots + x_1B^1 + x_0B^0 + x_{-1}B^{-1} + \ldots + x_{-p}B^{-p}$$

where B, the indices, and the exponents are expressed in base ten.

Digit x_i of *rank* i is said to have a *weight* of B^i and we can write:

$$X_{(\text{ten})} = \sum_{i=-p}^{n-1} x_i B^i$$

The use of data-processing machines has brought bases of ten, two, and sixteen into use.

— *base ten* is the habitual operating base of the user.

— *base two* is the operating base of the machine on account of the technology used. Remember that the binary digit is called a *bit*.

— *base sixteen* is a convenient means of coding numbers expressed in base two, since $16 = 2^4$.

In practice the following are found in these machines:

—*binary coded decimal* (BCD) where:

0 is coded by the binary combination 0000
1 is coded by the binary combination 0001
2 is coded by the binary combination 0010
3 is coded by the binary combination 0011
4 is coded by the binary combination 0100
5 is coded by the binary combination 0101
6 is coded by the binary combination 0110

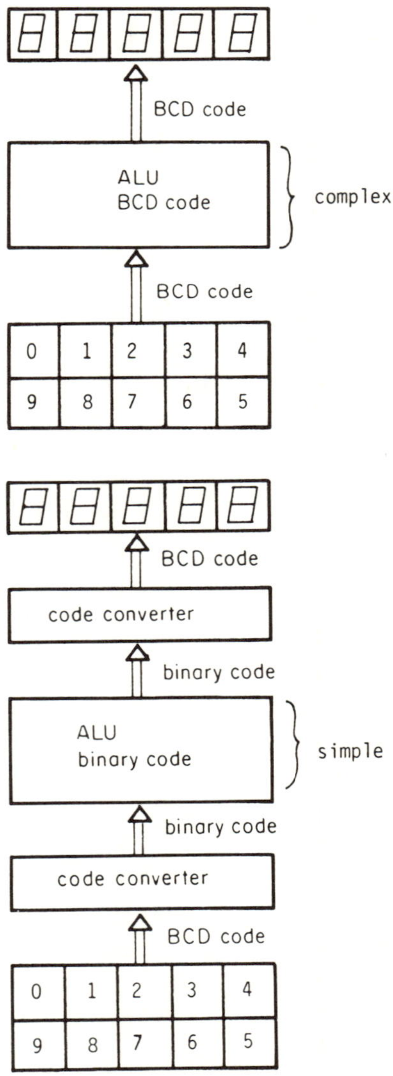

Figure 5.1

7 is coded by the binary combination 0111
8 is coded by the binary combination 1000
9 is coded by the binary combination 1001

— a code which can be called *binary coded hexadecimal* (BCH) where the figures 0 to 9 are coded as previously, to which these figures are added:

A coded by the binary combination 1010
B coded by the binary combination 1011
C coded by the binary combination 1100
D coded by the binary combination 1101
E coded by the binary combination 1110
F coded by the binary combination 1111

It is possible to carry out arithmetic operations in the two codes. Figure 5.1 shows the necessity to convert data from one code to the other.

Unsigned representation of numbers

Fixed point: Any number is represented by a word of n bits, of which i bits are assigned to the integral part I and f bits to the fractional part F, where $i+f=n$.

The number is written $N=I \cdot F$ (Figure 5.2).

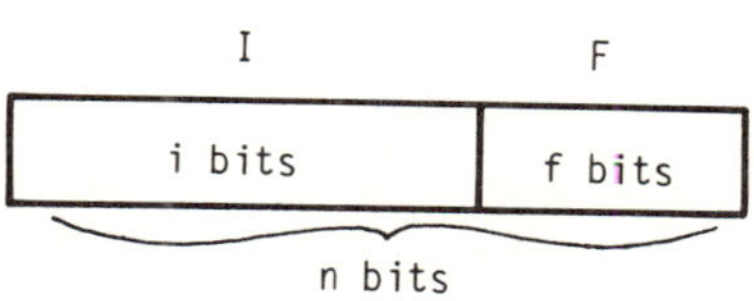

Figure 5.2

Example: $N=10110.011$, hence $i=5$, $f=3$, $n=8$.
$I=10110$ or 22 in base ten,
$F=0.011$ or 0.375 in base ten,
Hence $N=22.375$

Notice, for these values of i and f:
— that 0 is written 00000.000;
— that the smallest nonzero value, called the *quantum*, is 00000.001 or 0.125;
— that the highest value is 11111.111 or 31.875.

For a given n, if f is small, the quantum is large and the pecision reduced, but large values are reached. The converse applies if f is large.

Example:

$$n=8, f=0 \Rightarrow i=8: \text{quantum}=1, \quad N_{\max}=255.$$
$$n=8, f=8 \Rightarrow i=0: \text{quantum}=1/256, \quad N_{\max}=255/256.$$

Floating point: Any number can be written in the form:

$$N = M \times B^E$$

where M is called the *mantissa*, B the *base* (generally two, ten, or sixteen), and E the *exponent*.

B is implicit; only E and M are expressed in a binary code where m bits are assigned to the mantissa and e bits to the exponent, with $m + e = n$ (Figure 5.3).

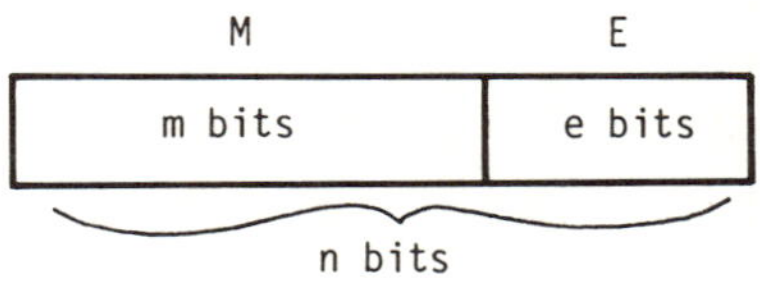

Figure 5.3

Example: Reconsider the previous example. The base ten number 22.375 is written 10110.011 in fixed point representation. The number can also be written:

$$1011.0011 \times 2^1$$
$$101.10011 \times 2^2$$
$$10.110011 \times 2^3$$
$$1.0110011 \times 2^4$$
$$0.10110011 \times 2^5$$
$$0.010110011 \times 2^6$$
etc.

Of all the possible representations, that in *which the first digit after the point is 1* is called *normalized representation*, in this case 0.10110011×2^5.

The format of the number is given in Figure 5.4.

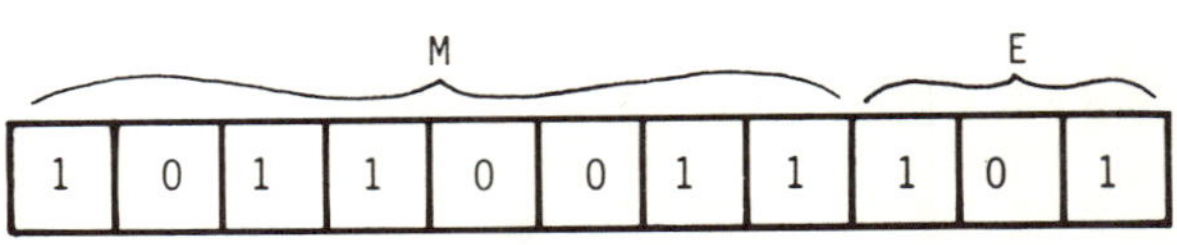

Figure 5.4

Notice that it is also possible to reverse the order (Figure 5.5).

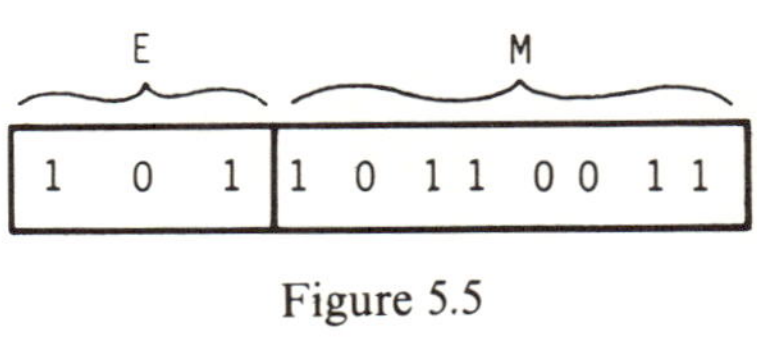

Figure 5.5

Floating point allows the representation of large numbers with a precision adjusted relatively. On the other hand, it facilitates management of numbers by the data-processing specialist.

Signed representation of numbers.

Two representations are used:

Modulus and sign: The unsigned representation is preceded by a bit called the *sign bit*, written sgn, equal to 0 for positive numbers and 1 for negative numbers.

This bit is not weighted by a power of the base; it is, therefore, treated differently from the bits of the number:

$$X = \text{sgn}\,X\; x_{n-1}x_{n-2}\ldots x_0 \cdot x_{-1}\ldots x_{-p}$$

for a fixed point representation, and

$$X = \text{sgn}\,M\; m_{n-1}m_0 \cdot B^{\text{sgn}\,E e_{q-1}\ldots e_0}$$

for a floating point representation.

However, modulus and sign representation leads to a complicated operation for addition and subtraction, so one tends to use the following representation in preference.

B's complement: Although this representation is valid for all bases B, we shall recall only those facts relative to $B = 2$.

In this representation, the digit of rank $(n-1)$ has a weight of -2^{n-1} instead of $+2^{n-1}$ in an unsigned representation.

Consider a binary representation $X = x_{n-1}x_{n-2}\ldots x_1 x_0$
An unsigned representation can be associated:

$$NX = x_{n-1}2^{n-1} + x_{n-2}2^{n-2} + \ldots + x_1 2^1 + x_0 2^0,$$

and a signed representation in 2's complement:

$$NX^* = -x_{n-1}2^{n-1} + x_{n-2}2^{n-2} + \ldots + x_1 2^1 + x_0 2^0,$$

from which

$$NX^* = -x_{n-1}2^n + NX.$$

Example: Let $X = 10100$, $n = 5$, $2^n = 32$, therefore, $NX = 20$ and $NX^* = -32 + 20 = -12$.

Comment: In the case of floating point representation, the mantissa and/or the exponent can be expressed in 2's complement form.

5.2 STATEMENT OF EXERCISES AND PROBLEMS

Preliminary comment: In many problems related to arithmetic processing of binary data, occasional difficulties can be overcome by reference to base ten.

5.E.1 Parallel binary addition with look-ahead carry

This technique is used order to avoid the inconvenience of serial carry propagation.

Consider the block diagram of the adder in Figure 5.6. The result is established when the carry r_4 is evaluated.

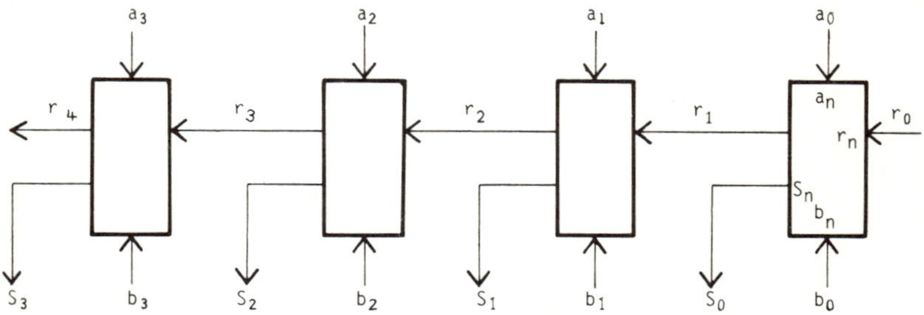

Figure 5.6

1. Show that r_4 depends on all the input variables. Hence deduce a new configuration for the adder.

2. Establish the relations giving r_1, r_2, r_3, r_4 as functions of the input variables. Put $G_n = a_n \cdot b_n$, the generation term, and $P_n = a_n + b_n$, the propagation term of stage n.

3. Recall the relations giving S_0, S_1, S_2, S_3 as functions of the input variables and carries. Show that the sum S_n of rank n is also a function of G_n, P_n, r_n.

4. Give the diagram of such an adder for $n = 4$ stages.

5. Knowing that the propagation delay of one logic stage is 15 ns, determine the calculation time T and show that it is independent of n. Compare with the results of problem *3.E.1*.

5.E.2 Operation of the arithmetic and logic unit (ALU) 74181

The logic diagram of the unit is given in Figure 5.7 and the table of propagation times, t_{pd}, Figure 5.8.

1. Give a model of the component, grouping the different inputs and outputs rationally.

This component is realized with basic logic operations; we can, therefore analyse its logic function. However, it realizes binary arithmetic operations; we can, therefore, analyse its operation equally from this point of view.

5.E.2.1 Logical aspect

2. Extract from Figure 5.7 the arrangement for obtaining the generation and propagation terms G_i and P_i for stage i. Write expressions for G_i and P_i.

3. What do these expressions become when the circuit realizes:

— an addition $S_3S_2S_1S_0 = 1001$;

— a subtraction $S_3S_2S_1S_0 = 0110$?

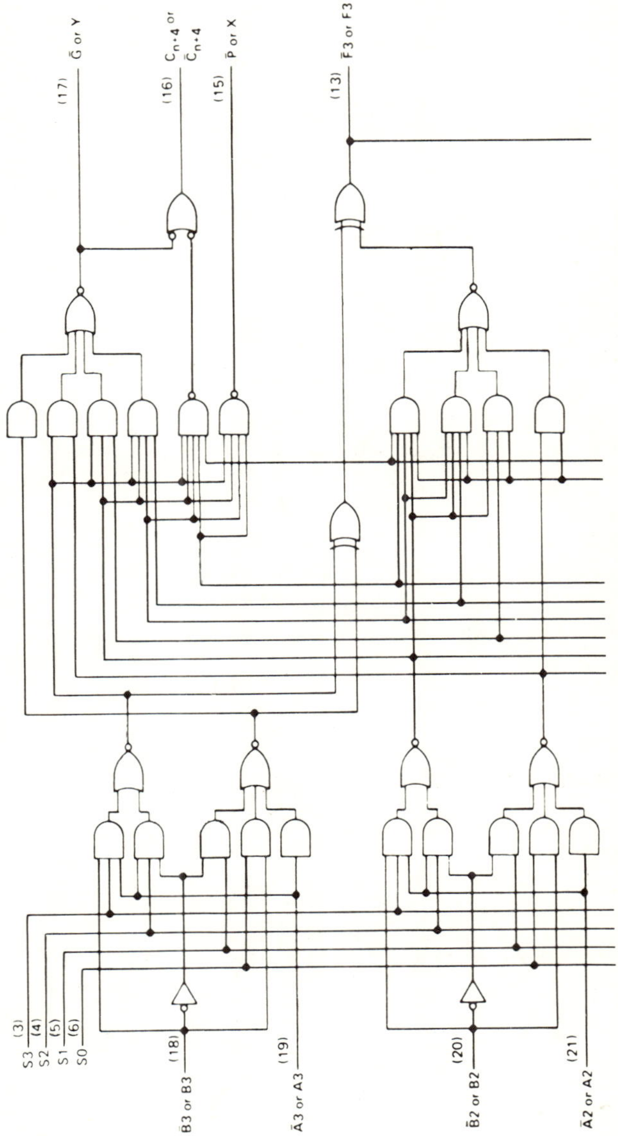
S3 (3)
S2 (4)
S1 (5)
S0 (6)
B̄3 or B3 (18)
Ā3 or A3 (19)
B̄2 or B2 (20)
Ā2 or A2 (21)
(17) Ḡ or Y
(16) C_{n+4} or $\bar{C}_{n+4}$
(15) P̄ or X
(13) F̄3 or F3

Figure 5.7

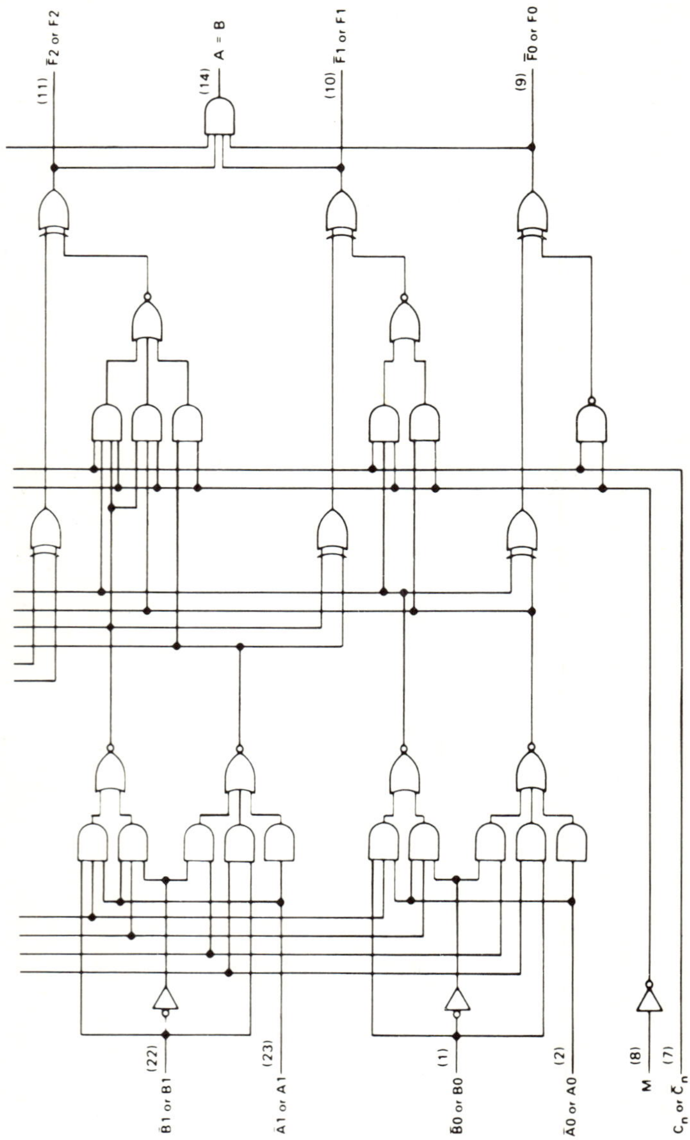

Figure 5.7 (*Contd.*)

PARAMETER	FROM (INPUT)	TO (OUTPUT)	TEST CONDITIONS	MIN	TYP	MAX	UNIT
t_{PLH}	C_n	C_{n+4}			12	18	ns
t_{PHL}					13	19	
t_{PLH}	Any $\bar{A}$ or $\bar{B}$	C_{n+4}	M = 0V,S0 = S3 = 4.5V.		28	43	ns
t_{PHL}			S1 = S2 = 0V($\overline{\text{SUM}}$ model)		27	41	
t_{PLH}	Any $\bar{A}$ or $\bar{B}$	C_{n+4}	M = 0V, S0 = S3 = 0V,		35	50	ns
t_{PHL}			S1 = S2 = 4.5V($\overline{\text{DIFF}}$ model)		33	50	
t_{PLH}	C_n	Any $\bar{F}$	M = 0V		13	19	ns
t_{PHL}			($\overline{\text{SUM}}$ or $\overline{\text{DIFF}}$ mode)		12	18	
t_{PLH}	Any $\bar{A}$ or $\bar{B}$	$\bar{G}$	M = 0V, S0 = S3 = 4.5V,		13	19	ns
t_{PHL}			S1 = S2 = 0V($\overline{\text{SUM}}$ mode)		13	19	
t_{PLH}	Any $\bar{A}$ or $\bar{B}$	$\bar{G}$	M = 0V, S0 = S3 = 0V,		17	25	ns
t_{PLH}			S1 = S2 = 4.5V($\overline{\text{DIFF}}$ mode)		17	25	
t_{PLH}	Any $\bar{A}$ or $\bar{B}$	$\bar{P}$	M = 0V, S0 = S3 = 4.5V,		13	19	ns
t_{PLH}			S1 = S2 = 0V($\overline{\text{SUM}}$mode)		17	25	

'PLH	Any $\bar{A}$ or $\bar{B}$	$\bar{P}$	M = 0 V, S0 = S3 = 0 V,	17	25	ns
'PHL			S1 = S2 = 4.5 V($\overline{DIFF}$ mode)	17	25	
'PLH	$\bar{A}_i$ or $\bar{B}_i$	$\bar{F}_i$	M = 0 V, S0 = S3 = 4.5 V,	28	42	ns
'PHL			S1 = S2 = 0 V($\overline{SUM}$ mode)	21	32	
'PLH	$\bar{A}_i$ or $\bar{B}_i$	$\bar{F}_i$	M = 0 V, S0 = S3 = 0 V,	32	48	ns
'PHL			S1 = S2 = 4.5 V($\overline{DIFF}$ mode)	23	34	
'PLH	$\bar{A}_i$ or $\bar{B}_i$	$\bar{F}_i$	M = 4.5 V(logic mode)	32	48	ns
'PHL				23	34	
'PLH	Any $\bar{A}$ or $\bar{B}$	A = B	M = 0 V, S0 = S3 = 0 V,	35	50	ns
'PHL			S1 = S2 = 4.5 V($\overline{DIFF}$ mode)	32	48	

'PLH = propagation delay time, low-to-high-level output
'PHL ≡ propagation delay time, high-to-low-level output

Figure 5.8 Dynamic characteristic of the 74181

4. Extract from Figure 5.7 a diagram of the realization of the carry C_i and the output F_i for stage i. Verify that the carry conforms to the theory set out in problem *5.E.1*.

5. For $M=0$, the circuit realizes arithmetic operations. For this case express $F_i=f(G_i, P_i, C_i)$. In the case of addition and substraction replace G_i and P_i by their value obtained in problem 3. Comment on the results.

Note: The manufacturer publishes a table of the arithmetic operations realized for the 16 combinations of $S_3S_2S_1S_0$. Apart from addition and subtraction, they are of not of any interest; some can even qualify as 'traditional'. We shall not study them.

6. For $M=1$, the circuit realises logical operations. Express $F_i=f(G_i, P_i, C_i)$. Replace G_i and P_i by their values obtained in problem 2 above.

7. Set up a table giving $F_i=f(A_i, B_i)$ for all combinations of $S_3S_2S_1S_0$, indicating the logical operation realized.

5.E.2.2 Arithmetic aspect

Consider the model of Figure 5.9 where (see Section 5.1) NA is associated with word A, NB is associated with word B, and NF is associated with word F.

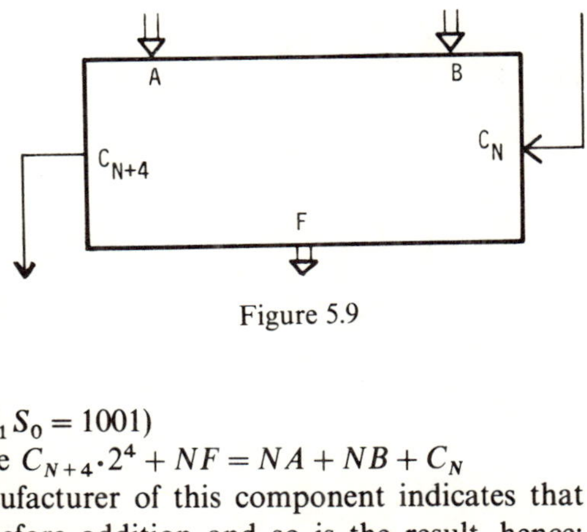

Figure 5.9

Addition $(S_3S_2S_1S_0=1001)$

8a. We require $C_{N+4}\cdot 2^4+NF=NA+NB+C_N$

Now the manufacturer of this component indicates that the operands are complemented before addition and so is the result, hence: $N\bar{A}$ is associated with $\bar{A}$, $N\bar{B}$ is associated with $\bar{B}$, and $N\bar{F}$ is associated with $\bar{F}$.

Show that $N\bar{A}=2^4-1-NA$.

Show that $NF=2^4(C_{N+4}-1)+NA+NB+\overline{C_N}$.

8b. Let us apply X and Y, active *at the high level* to A and B, and let S be the sum.

$$A=X \Rightarrow NA=NX$$
$$B=Y \Rightarrow NB=NY$$
$$F=S \Rightarrow NF=NS.$$

What happens to relation (1)? Discuss this result with respect to C_{N+4}.

8c. Let us apply $\bar{X}$ and $\bar{Y}$, active *at the low level.*

$$A = X \Rightarrow NA = 2^4 - 1 - NX$$
$$B = Y \Rightarrow NB = 2^4 - 1 - NY$$
$$F = S \Rightarrow NF = NS$$

Show that $N\bar{S} = -2^4 C_{N+4} + NX + NY + C_N$. Discuss with respect to C_{N+4}.

Subtraction ($S_3S_2S_1S_0 = 0110$)

9a. The manufacturer of the component indicates that in addition to the complementing indicated previously, the 74181 adds the complement of B instead of subtracting B.

Show that $-C_{N+4}2^4 + NF = N_A - N_B - C_N$, a classic result. (2)

9b. Repeat Questions 8b and 8c, adapting them for subtraction.

9c. Summarize the preceding study.

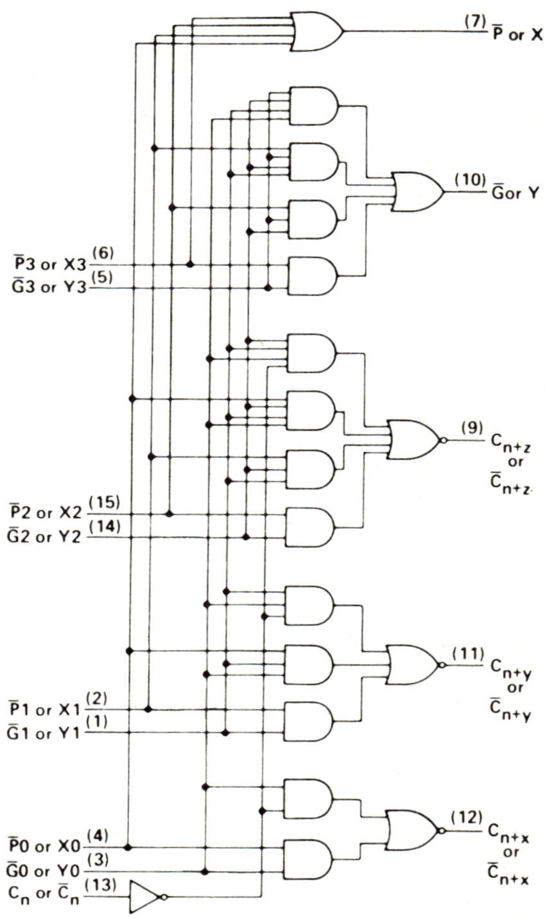

Figure 5.10 The maximum propagation time is 22 ns.

Connection of several components

10. Give the diagram of a 16-bit adder-subtracter. The data can be active high or low: a switch allows that choice. Also, the operation is selected with the help of bit OP:

$$OP = 1, \text{ addition} \qquad OP = 0, \text{ subtraction.}$$

5.E.3 Operation of the carry look-ahead circuit 74182

1. Considering the diagram of the 74181 or 14581B (Figure 5.7), write expressions for $\bar{P}$, $\bar{G}$, and C_{N+4}.

2. Considering the diagram of the 74182 or 14582B given in Figure 5.10, write expressions for C_{N+X}, C_{N+Y}, C_{N+Z}, $\bar{G}$, and $\bar{P}$.

3. Consider the arrangement of Figure 5.11. Using the table of propagation times, (Figure 5.8), calculate the time necessary for the execution of one addition-subtraction.

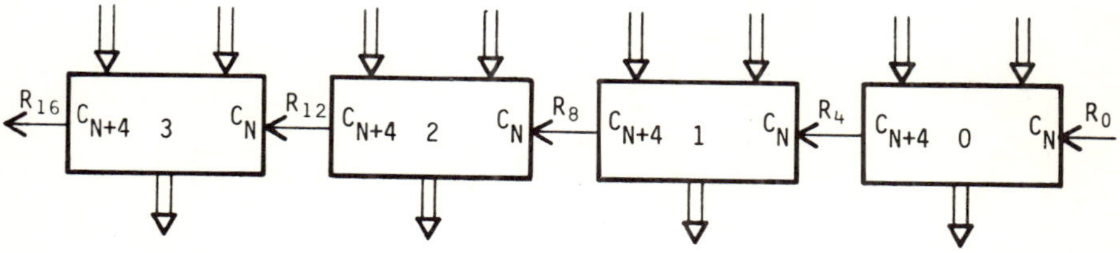

Figure 5.11

Now consider the arrangement of Figure 5.12.

4a. Show that $C_{N+X} = R_4$, $C_{N+Y} = R_8$, $C_{N+Z} = R_{12}$.

4b. Why are the $\bar{G}$ and $\bar{P}$ outputs of 74181 or 14581B *number 3* not connected to the $\bar{G}_3$ and $\bar{P}_3$ inputs of the 74182 or 14582B?

4c. Give the diagram of a 48-bit ALU, with carry look-ahead.

4d. Using the table of propagation times, (Figure 5.8), calculate the execution time of an addition-subtraction for an array of 16 bits.

4e. Compare with the arrangement of Figure 5.11.

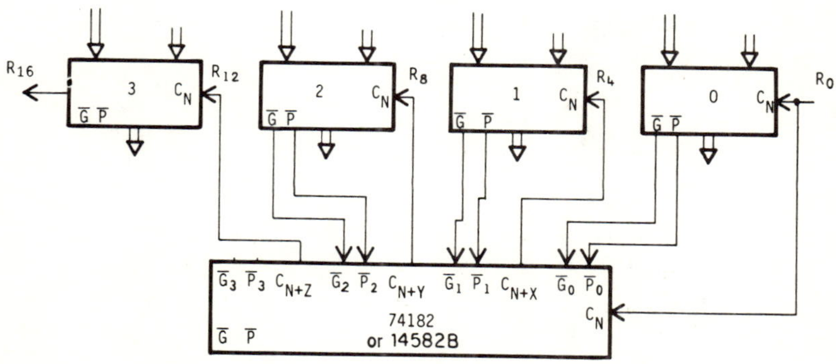

Figure 5.12

5.E.4 Decimal arithmetic in BCD code

5.E.4.1 Study of addition

Recall that, in this code, each character from 0 to 9 is coded by four bits having weights of 1, 2, 4, and 8. These groups of four bits are called *tetrads*. There are, therefore, ten tetrads from 0 to 9.

The six remaining combinations of these 4 bits, A = 1010 to F = 1111, are called *pseudo-tetrads* and are not part of the code.

The adder is modelled in Figure 5.13.

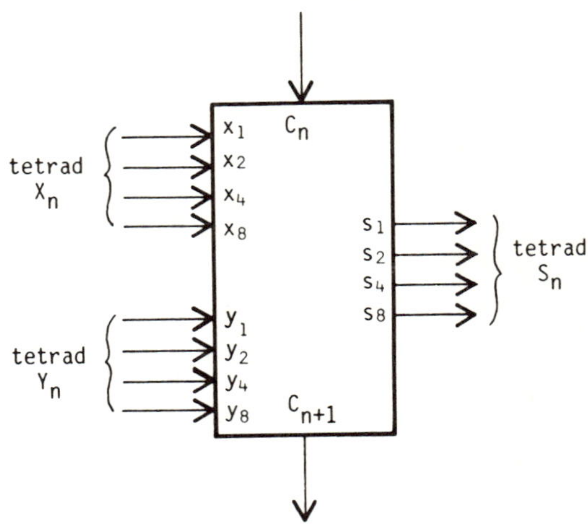

Figure 5.13

A first method of addition consists of treating the required function as a Boolean function. It is then necessary to synthesize 5 functions of 9 variables. We shall not consider this solution.

A second method consists of using binary adders which exist in the manufacturer's catalogues: 74283, etc. In this case, the result may be false; however, it can be corrected.

Example: 3 → 0011
6 → 0110

---

9 0 1001 → 9 *correct result*

5 → 0101
7 → 0111

---

12 0 1100 → C *false code*

In fact we should obtain $\underbrace{0001}_{1}\ \underbrace{0010}_{2}$.

$$\begin{array}{r} 9 \rightarrow 1001 \\ 9 \rightarrow 1001 \\ \hline \end{array}$$

$$18\ \ 1\ 0010 \rightarrow 12 \text{ \textit{false result}}$$

In fact we should obtain $\underbrace{0001}_{1}\ \underbrace{1000}_{8}$.

Consider all the cases of decimal addition:

$$C_{n+1} \times 10 \quad \text{plus} \quad S_n = X_n \quad \text{plus} \quad Y_n \quad \text{plus} \quad C_n.$$

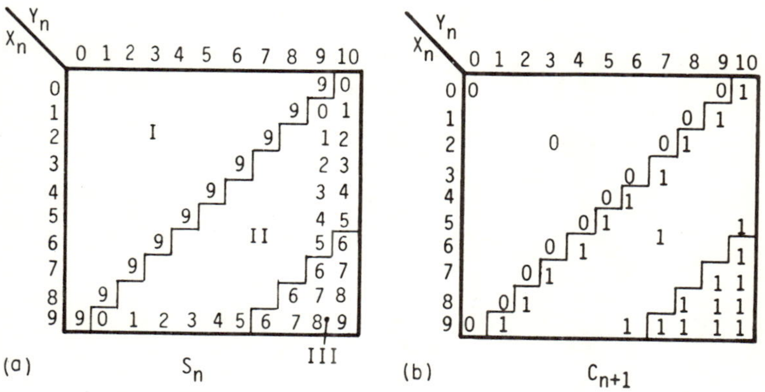

The column on the right takes account of a carry $C_n = 1$

Figure 5.14

The table of Figure 5.14 shows that if $X_n = 9$, $Y_n = 9$ (maximum values) and $C_n = 1$, we obtain $S_n = 9$ and $C_{n+1} = 1$

$$9 \quad \text{plus} \quad 9 \quad \text{plus} \quad 1 = 19$$

In fact the adder used works in binary code which can be 'seen' as hexadecimal coded binary:

$$C'_{n+1} \times 16 \,\text{plus}\, S'_n = X_n \,\text{plus}\, Y_n \,\text{plus}\, C_n.$$

1. Zone IV (see Figure 5.15) does not arise in our case. Why?

Notice in Figures 5.14 and 5.15 that zones I are identical, that the pseudo-tetrads of zone II must be corrected and a carry C_{n+1} generated, that the false results of zone III must be corrected and the existing carry retained.

2. What mathematical correction can be made in zone II? Justify.
3. What mathematical correction can be made in zone III? Justify.
4. The diagram of the corresponding principle is given in Figure 5.16. Give a detailed diagram. What is obtained on C_4 of ADD2?
5. Modify the previous diagram, by omitting the block (?) and exploiting C_4 of ADD2.

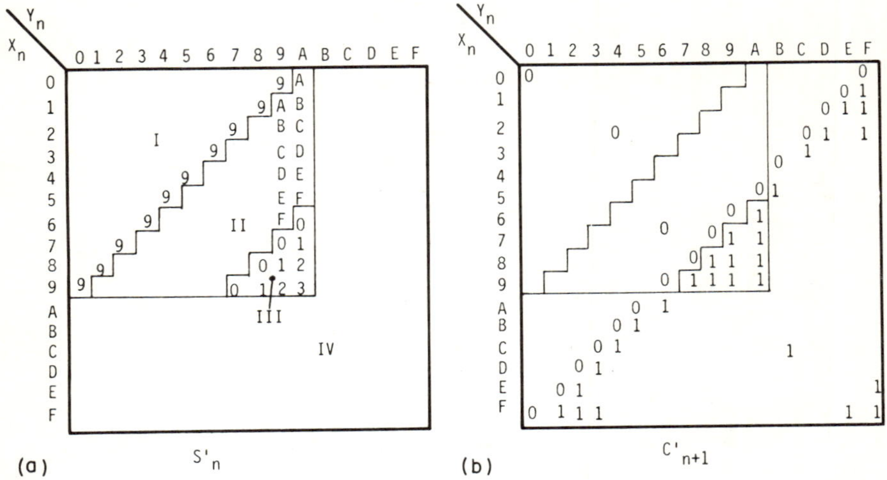

Figure 5.15

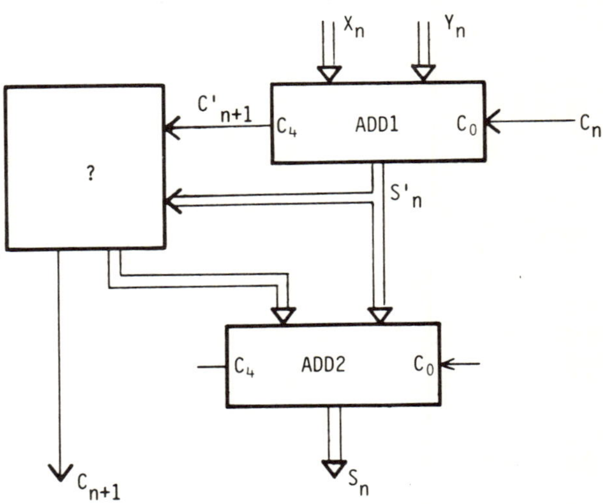

Figure 5.16

6. A third method consists of using an adder and treating the corrector as a Boolean function (Figure 5.17).

Express $C_{n+1}, S_{3n}, S_{2n}, S_{1n}, S_{0n}$ as functions of $C'_{n+1}, S'_{3n}, S'_{2n}, S'_{1n}, S'_{0n}$.

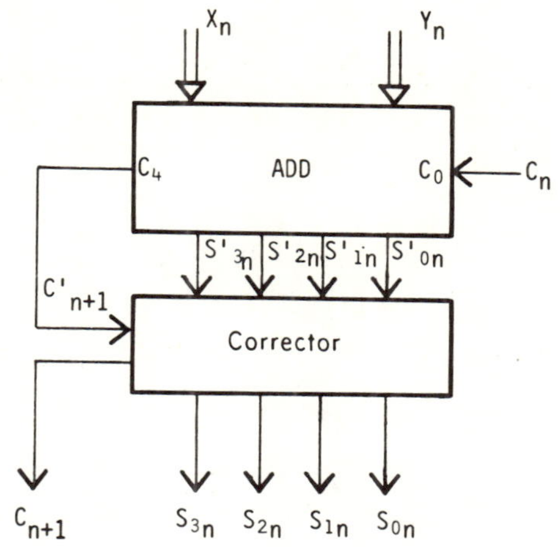

Figure 5.17

5.E.5 Addition and subtraction of numbers expressed as modulus and sign

Recall that each number is preceded by a sign bit (0 for positive numbers and 1 for negative numbers) and that binary subtraction is carried out in the order: larger modulus less the smaller modulus.

5.E.5.1 Addition

1. Taking all possibilities into account, draw a table representing the operation.
2. Devise an architecture which simplifies the solution.
3. Give a diagram, using TTL circuits, for a format of 8 bits including sign.

5.E.5.2 Subtraction

Repeat the previous questions.

5.E.5.3 Addition-subtraction

Give the diagram of an 8-bit adder-subtractor including sign, the operation required being chosen by the signal S/D, where:

$$S/D = 0 \Rightarrow \text{ADD}, \qquad S/D = 1 \Rightarrow \text{SUB}.$$

5.E.6 Addition and subtraction of numbers expressed in 2's complement form

Recall that the sign bit is also placed in front, but it is weighted by -2^{n-1} instead of $+2^{n-1}$ for an unsigned number.

5.E.6.1 Addition

1. Starting with the relations relating to B's complement on page 127, show that a binary adder is suitable for the operation, having regard for an overflow condition, ovr, which will be explained.
2. Express:

$$\text{ovr} = f(x_{n-1}, y_{n-1}, r_{n-1}, C_n),$$
$$\text{ovr} = g(C_{n-1}, C_n),$$
$$\text{ovr} = h(x_{n-1}, y_{n-1}, r_{n-1}),$$

where f, g, and h are Boolean functions of the following binary variables:

$x_{n-1}, y_{n-1}, r_{n-1}$, sign bits of the operands and the result,
C_{n-1}, C_n, carries 'in the sign' and 'beyond the sign'.

5.E.6.2 Subtraction

1. Show that a binary adder is suitable, the subtraction of an operand being replaced by the addition of a quantity to be specified.
2. Deduce a scheme which realises addition and subtraction, the operation to be executed being chosen by the signal OP:

$$\text{OP} = 0 \Rightarrow \text{ADD},$$
$$\text{OP} = 1 \Rightarrow \text{SUB}.$$

5.E.7 Ternary matrix arithmetic

Another way of coding a number in binary consists of using a word of B bits (one bit for each symbol of the base) for the *counting base B*. When this symbol occurs, the corresponding bit is 1.

Example: $B = 10$

	9	8	7	6	5	4	3	2	1	0
	a_9	a_8	a_7	a_6	a_5	a_4	a_3	a_2	a_1	a_0
0	0	0	0	0	0	0	0	0	0	1
1	0	0	0	0	0	0	0	0	1	0
⋮										
5	0	0	0	0	1	0	0	0	0	0
⋮										
9	0	0	0	0	1	0	0	0	0	0

In the following, we shall consider base 3, in order to reduce the complexity of the circuits, $B = 3$.

	2	1	0
	a_2	a_1	a_0
0	0	0	1
1	0	1	0
2	1	0	0

Multi-digit numbers are, therefore, formed from binary triplets, each one coding a digit of the number.

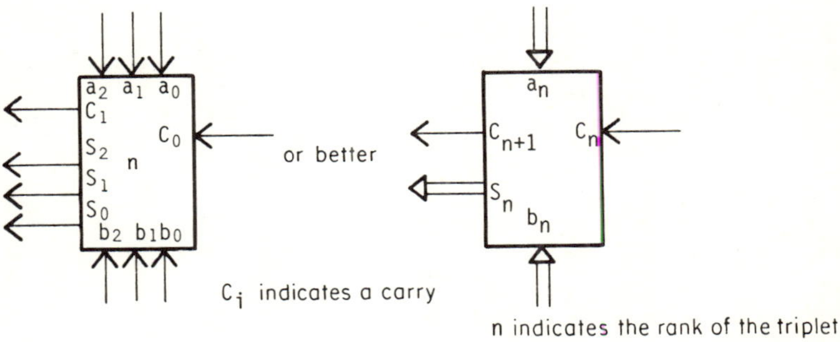

Figure 5.18

Example: $(212)_3$ is coded as 100 010 100

We shall study addition modulo 3 in this code. The corresponding adder will be modelled according to Figure 5.18.

1. Complete the modulo 3 addition tables of Figure 5.19, without taking account of the incoming carry C_n.

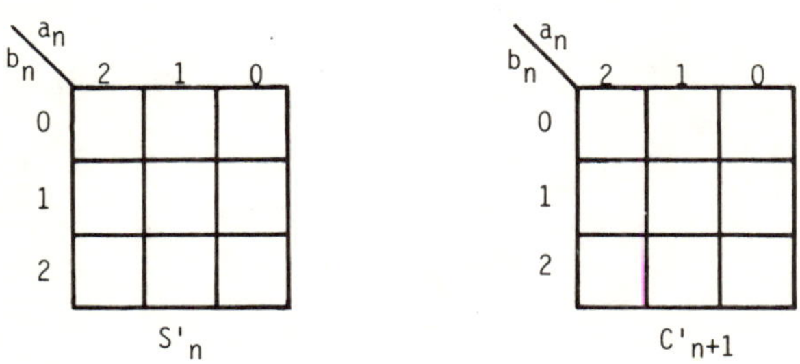

Figure 5.19

Carry out the following additions in the code described previously:

$$\begin{array}{r} 010 \\ +001 \\ \hline \end{array} \qquad \begin{array}{r} 100 \\ +010 \\ \hline \end{array} \qquad \begin{array}{r} 100 \\ +100 \\ \hline \end{array}$$

↑ carry | sum ↑

Note: The + sign represents the arithmetic sum.

2a. Show that a matrix organization enables this adder to be synthesized. Make use of the tables in problem 1 above.

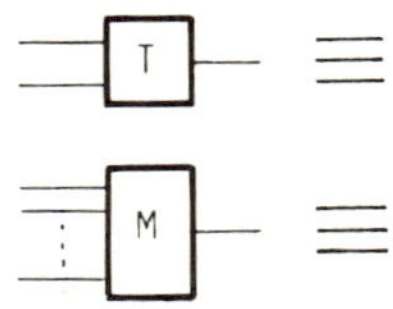

2b. Clarify the operations of Figure 5.20 represented above, using the binary operators which you know.

On Figure 5.20, join the inputs of the operators **M** to the correct outputs of the operators T.

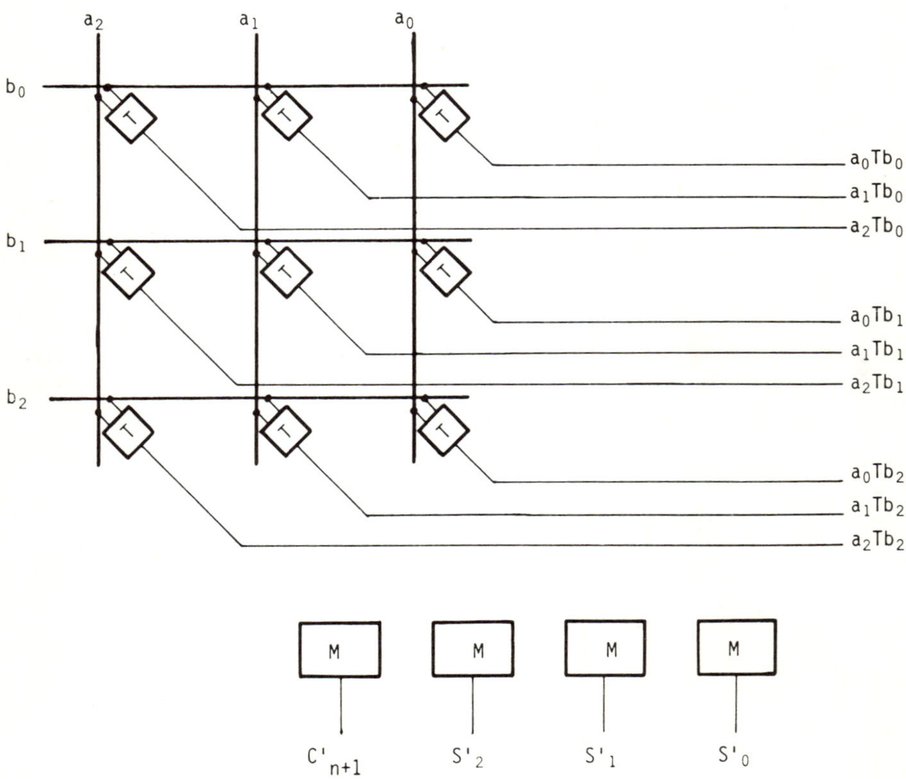

Figure 5.20

2c. Clarify the relations:

$$S'_0 = f_0(a_2, a_1, a_0, b_2, b_1, b_0) \qquad S'_1 = f_1(a_2, a_1, a_0, b_2, b_1, b_0)$$

$$S'_2 = f_2(a_2, a_1, a_0, b_2, b_1, b_0) \qquad C'_{n+1} = f_3(a_2, a_1, a_0, b_2, b_1, b_0)$$

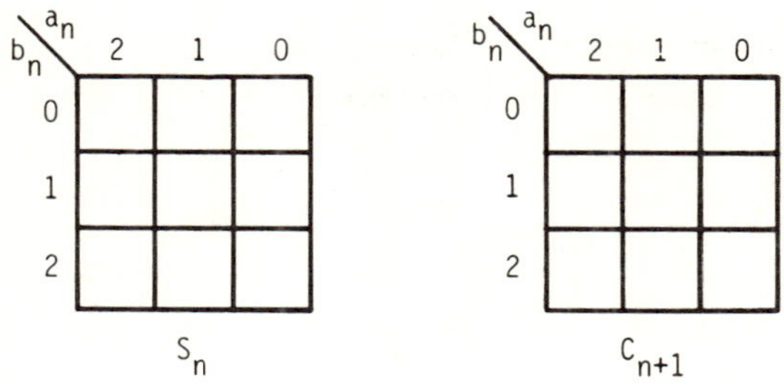

Figure 5.21

3. Complete the tables of modulo 3 addition in Figure 5.21, taking account of the input carry $C_n = 1$. (If $C_n = 0$, the case of problem 1 applies).

Carry out the following additions in the code described previously:

```
  0 1 0 ← Cn          0 1 0 ← Cn          0 1 0 ← Cn
  ----------          ----------          ----------
  0 1 0               1 0 0               1 0 0
+ 0 0 1             + 0 1 0             + 1 0 0
```

```
  0 0 1 ← Cn          0 1 0 ← Cn          0 0 1 ← Cn
  ----------          ----------          ----------
  0 1 0               0 0 1               0 0 1
+ 0 1 0             + 1 0 0             + 0 0 1
```

Carry out the following addition, presented in base 3 and translate it to the code studied. (Carries will appear.)

```
          0 ←   Carries
  ---------
  2 1 0 2
+ 1 2 1 1
```

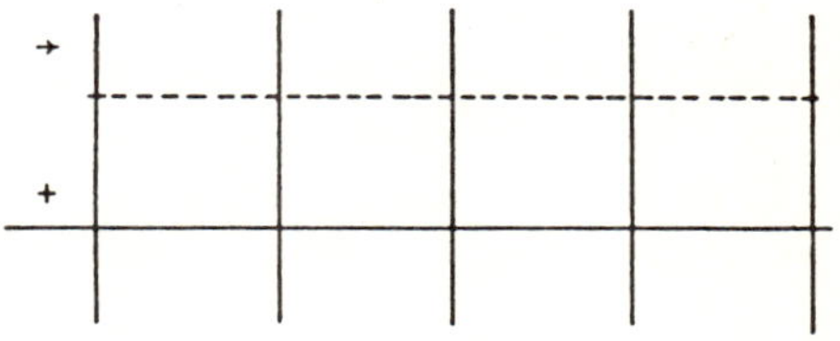

Note: The + sign represents the arithmetic sum.

4. Complete the diagram of Figure 5.22, taking account of the carry C_n. (The operators T and M will be replaced by their equivalents found in problem 2.)

5. Clarify the relations:

$$S_0 = g_0(a_2, a_1, a_0, b_2, b_1, b_0, C_n) \qquad S_2 = g_2(a_2, a_1, a_0, b_2, b_1, b_0, C_n)$$
$$S_1 = g_1(a_2, a_1, a_0, b_2, b_1, b_0, C_n) \qquad C_{n+1} = g_3(a_2, a_1, a_0, b_2, b_1, b_0, C_n).$$

6. *Interconnection*

6a. Give an interconnection of four modules (modelled according to Figure 5.18) which allows 4-digit ternary numbers to be treated with serial carry propagation.

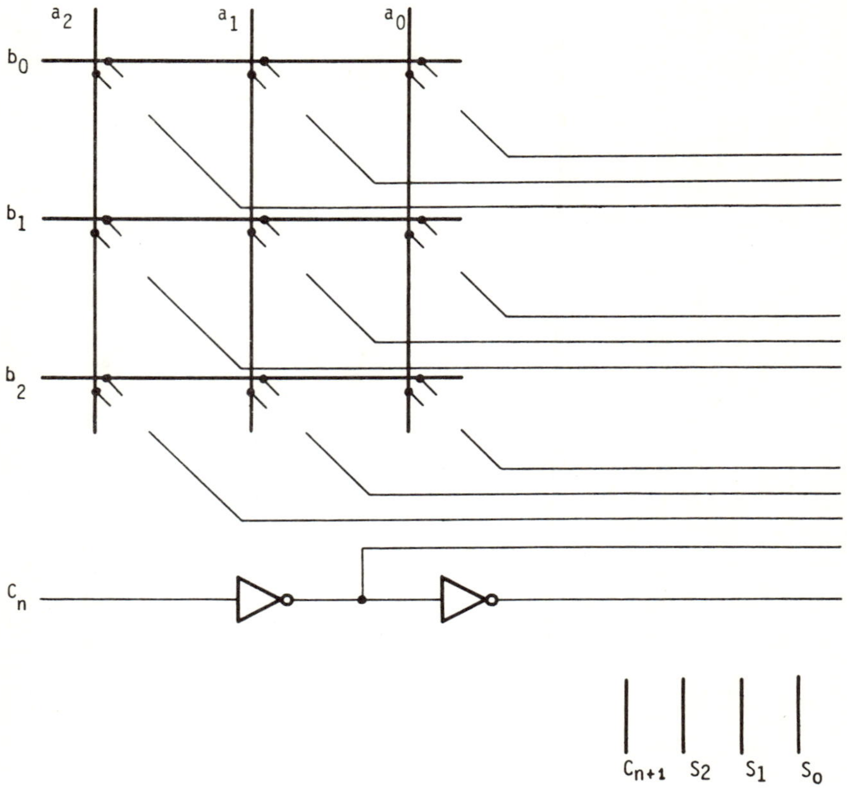

Figure 5.22

6b. Knowing that each basic circuit of an adder has a propagation time of 10 ns, calculate the calculation time of the preceding arrangement.

6c. In order to improve this time, it is decided to use the *look-ahead carry* technique. Show that C_{n+1} can be written in the form:

$$C_{n+1} = \delta_n + \gamma_n C_n. \text{ Evaluate } \delta_n \text{ and } \gamma_n.$$

6d. Clarify the relations:

$$C_1 = h_1(\delta_0, \gamma_0, C_0) \qquad C_3 = h_3(\delta_0, \gamma_0, C_0, \delta_1, \gamma_1, \delta_2, \gamma_2)$$
$$C_2 = h_2(\delta_0, \gamma_0, C_0, \delta_1, \gamma_1) \qquad C_4 = h_4(\delta_0, \gamma_0, C_0, \delta_1, \gamma_1, \delta_2, \gamma_2, \delta_3, \gamma_3)$$

Use the results of problem 5.E.1.

6e. Show that the 74182 (whose characteristics are given in Figure 5.10) is suitable for this function.

6f. Show that by adding two outputs to each of the adders, it is possible to use the 74182 to produce a 4-digit ternary adder with look-ahead carry. Give the diagram of this arrangement and indicate the corresponding calculation time.

5.3 SOLUTIONS

5.S.1 Parallel binary addition with look-ahead carry

1. Successive carrying forward allows the following relation to be established:

$$r_4 = f_4\{a_3, b_3, f_3[a_2, b_2, f_2[a_1, b_1, f_1(a_0, b_0, r_0)]]\}$$

which shows clearly that all the input variables appear, hence the arrangement of Figure 5.23.

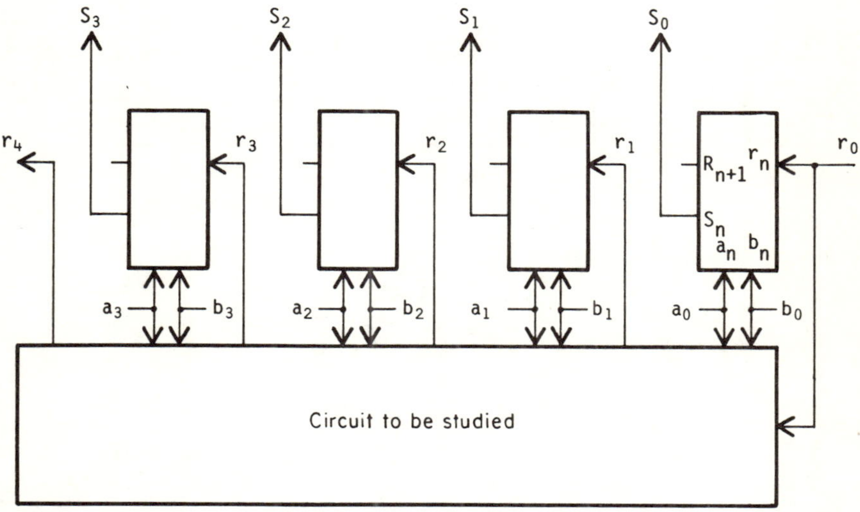

Figure 5.23

We see that the adders can be simplified since the function R_{n+1} is no longer used.

2. Recall that $R_{n+1} = a_n \cdot b_n + r_n \cdot (a_n + b_n)$, hence $R_{n+1} = G_n + r_n \cdot P_n$, from which:

$$r_1 = G_0 + r_0 \cdot P_0$$
$$r_2 = G_1 + r_1 \cdot P_1 = G_1 + P_1 \cdot G_0 + P_1 \cdot P_0 \cdot r_0$$
$$r_3 = G_2 + r_2 \cdot P_2 = G_2 + P_2 \cdot G_1 + P_2 \cdot P_1 \cdot G_0 + P_2 \cdot P_1 \cdot P_0 \cdot r_0$$
$$r_4 = G_3 + r_3 \cdot P_3 = G_3 + P_3 \cdot G_2 + P_3 \cdot P_2 \cdot G_1 + P_3 \cdot P_2 \cdot P_1 \cdot G_0 + P_3 \cdot P_2 \cdot P_1 \cdot P_0 \cdot r_0$$

3. $S_0 = a_0 \oplus b_0 \oplus r_0 \qquad S_2 = a_2 \oplus b_2 \oplus r_2$

$S_1 = a_1 \oplus b_1 \oplus r_1 \qquad S_3 = a_3 \oplus b_3 \oplus r_3$

An algebraic demonstration of the required relation could be given, but a graphical demonstration is preferred:

The K maps relating to $S_n = a_n \oplus b_n \oplus r_n$, $G_n = a_n \cdot b_n$, and $P_n = a_n + b_n$ are given in Figures 5.24, 5.25, and 5.26 respectively.

r_n \ $a_n b_n$	00	01	11	10
0	0	1	0	1
1	1	0	1	0

Figure 5.24

$a_n b_n$	00	01	11	10
	0	0	1	0

Figure 5.25

$a_n b_n$	00	01	11	10
	0	1	1	1

Figure 5.26

Hence we can write the graphical equality represented by Figure 5.27.

r_n \ $a_n b_n$	00	01	11	10
0	0	1	0	1
1	1	0	1	0

$\equiv$

r_n \ $a_n b_n$	00	01	11	10
0	0	1	1	1
1	0	0	1	0

$\bullet$

r_n \ $a_n b_n$	00	01	11	10
0	-	1	0	1
1	-	-	1	-

$+$

r_n \ $a_n b_n$	00	01	11	10
0	0	-	0	-
1	1	0	-	0

Figure 5.27

From this equality we deduce the relation:

$$
\begin{aligned}
S_n &= (\bar{r}_n\cdot P_n + r_n\cdot G_n)\cdot(r_n + \bar{a}_n + \bar{b}_n) + r_n\cdot\bar{a}_n\cdot\bar{b}_n \\
&= (\bar{r}_n\cdot P_n + r_n\cdot G_n)\cdot(r_n + \bar{G}_n) + r_n\cdot\bar{P}_n \\
&= r_n\cdot G_n + \bar{r}_n\cdot\bar{G}_n\cdot P_n + r_n\cdot\bar{P}_n \\
&= r_n\cdot G_n\cdot(P_n + \bar{P}_n) + \bar{r}_n\cdot\bar{G}_n\cdot P_n + r_n\cdot\bar{P}_n\cdot(G_n + \bar{G}_n) \\
&= r_n\cdot G_n\cdot P_n + \underline{r_n\cdot G_n\cdot\bar{P}_n} + \bar{r}_n\cdot\bar{G}_n\cdot P_n + r_n\cdot\bar{G}_n\cdot\bar{P}_n\cdot + r_n\cdot G_n\cdot P_n
\end{aligned}
$$

This expression is close to $S_n = G_n \oplus P_n \oplus r_n$, since:

$$a_n \oplus b_n \oplus r_n = r_n\cdot G_n\cdot P_n + \underline{\bar{r}_n\cdot G_n\cdot\bar{P}_n} + \bar{r}_n\cdot\bar{G}_n\cdot P_n + r_n\cdot\bar{G}_n\cdot\bar{P}_n.$$

The two expressions will not be identical unless the two underlined terms are identical, that is if $G_n\cdot\bar{P}_n = 0$.

Now, $G_n\cdot\bar{P}_n = a_n\cdot b_n\cdot\overline{(a_n + b_n)} = a_n\cdot b_n\cdot\bar{a}_n\cdot\bar{b}_n = 0$.

Hence $S_n = G_n \oplus P_n \oplus r_n$.

4. The diagram corresponding to such an adder is given in Figure 5.28. Also given are the diagram (Figure 5.29) of the 7483A TTL circuit (and 74283, alternative numbering) which, with some variations, closely corresponds to it; and (Figure 5.30) that of the CMOS MC14008B where only the final carry C_{out} is realized. According to the theory which has just been studied, it propagates the other carries serially.

5. It is easy to see that, for any number of stages, there are four levels of basic logic, hence $T = 15\,\text{ns} \times 4 \neq 60\,\text{ns}$.

Notice that if n is large, the complexity of the circuit increases greatly.

5.S.2 Operation of the arithmetic and logic unit (ALU) 74181 or 14581B

1. The model is given in Figure 5.31.

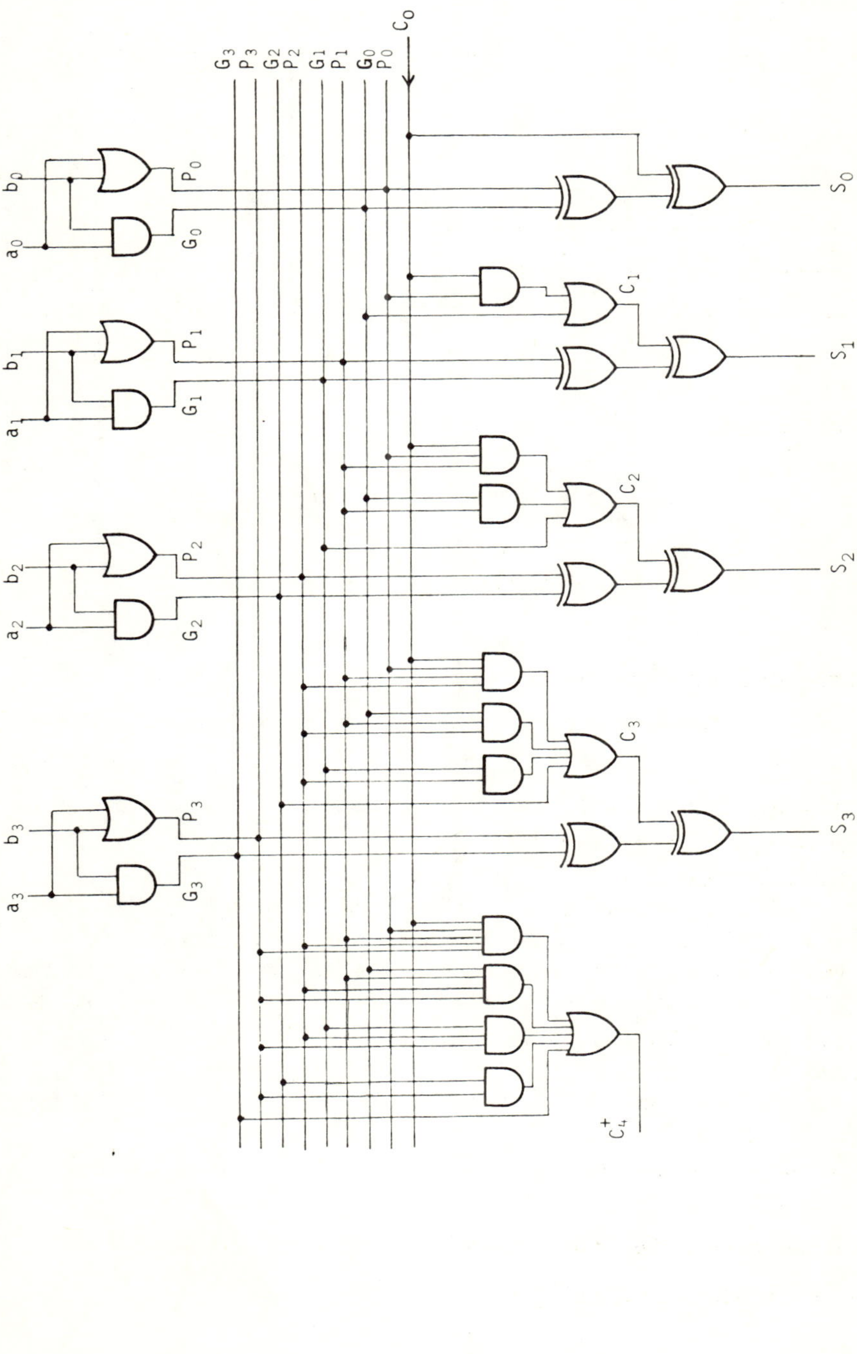

b_0
a_0
P_0
G_0
b_1
a_1
P_1
G_1
b_2
a_2
P_2
G_2
b_3
a_3
P_3
G_3
G_3 P_3 G_2 P_2 G_1 P_1 G_0 P_0
C_0
C_1
C_2
C_3
C_4^+
S_0
S_1
S_2
S_3

Figure 5.28

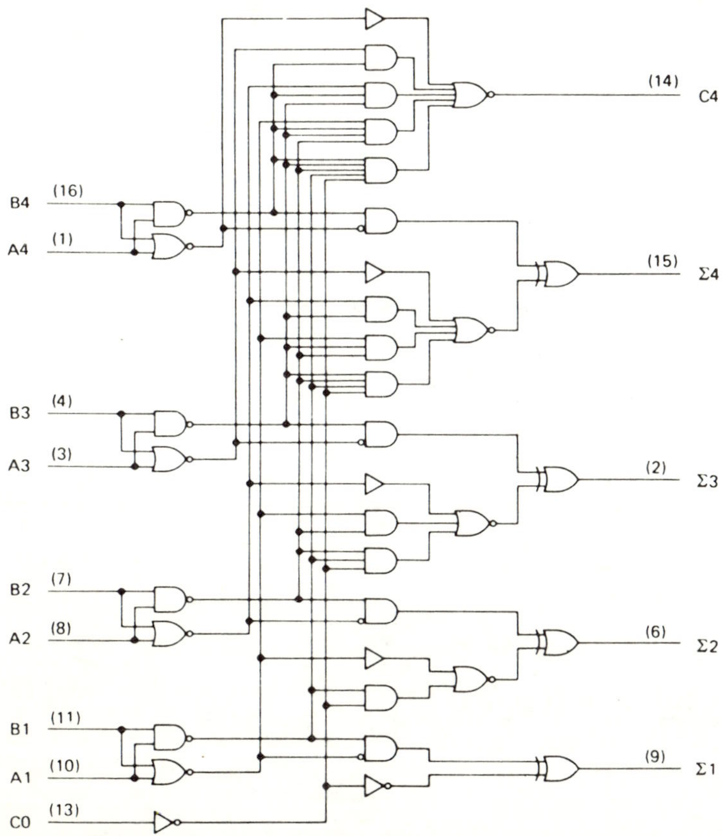

Figure 5.29

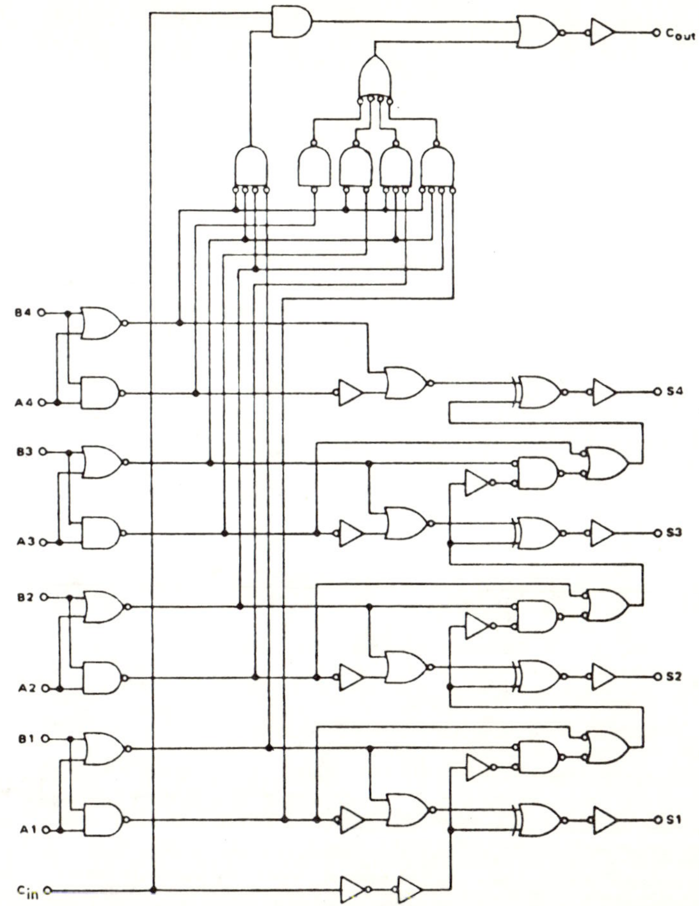

Figure 5.30

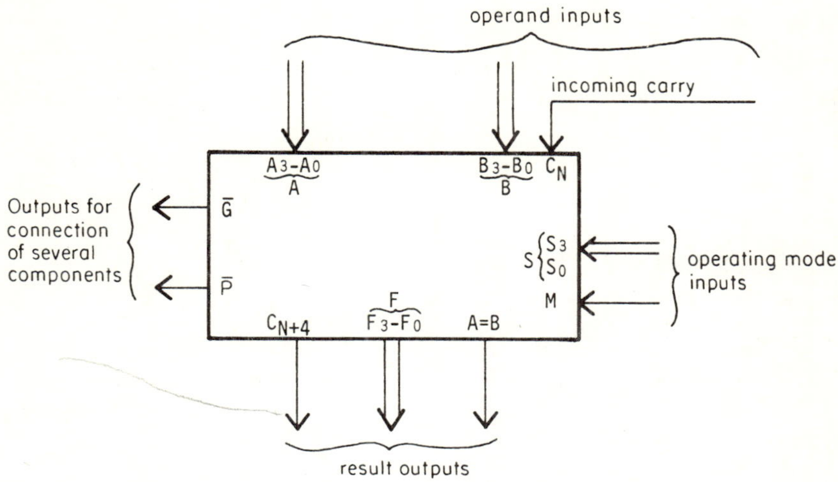

Figure 5.31

5.S.2.1 Logical aspect

2. The diagram is given in Figure 5.32.

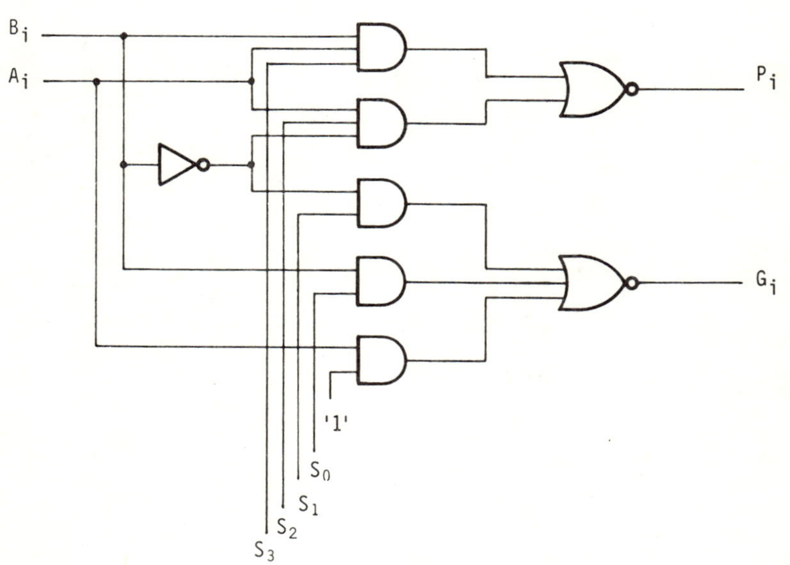

Figure 5.32

For $i \in [0, 1, 2, 3]$, we have $P_i = \overline{S_3 \cdot A_i \cdot B_i \cdot S_2 \cdot A_i \cdot \bar{B}_i}$ and $G_i = \overline{S_1 \cdot \bar{B}_i + S_0 \cdot B_i + A_i}$.

3. In the case of an addition: $S = 1001$

$$P_i = \overline{A_i \cdot B_i} = \bar{A}_i + \bar{B}_i, \quad G_i = \overline{A_i + B_i} = \bar{A}_i \cdot \bar{B}_i.$$

In the case of a subtraction: $S = 0110$

$$P_i = \overline{A_i \cdot \overline{B_i}} = \overline{A_i} + B_i, \quad G_i = \overline{A_i + \overline{B_i}} = \overline{A_i} \cdot B_i.$$

4. The outline diagram is given in Figure 5.33.

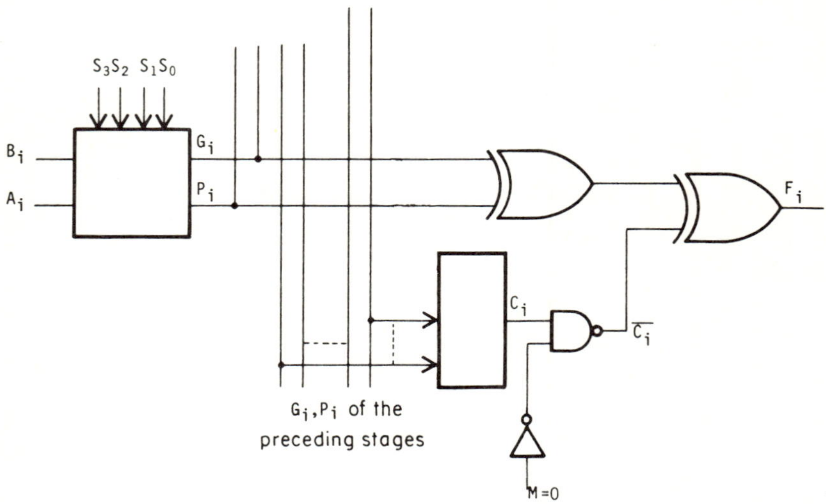

Figure 5.33

5. For $M = 0$, we have $F_i = G_i \oplus P_i \oplus \overline{C_i}$.
In the case of addition: $F_i = \overline{A_i + B_i} \oplus \overline{A_i \cdot B_i} \oplus \overline{C_i}$.

Now $x \oplus y \oplus z = (x \oplus y) \oplus z = (\bar{x} \cdot y + x \cdot \bar{y}) \oplus z$, hence:

$$F_i = (A_i + B_i) \cdot \overline{(A_i \cdot B_i)} + \overline{(A_i + B_i)} \cdot A_i \cdot B_i \oplus \overline{C_i}$$
$$= (A_i + B_i) \cdot (\overline{A_i} + \overline{B_i}) + \overline{A_i} \cdot \overline{B_i} \cdot A_i \cdot B_i \oplus \overline{C_i}.$$

Now $(A_i + B_i) \cdot (\overline{A_i} + \overline{B_i}) = A_i \oplus B_i$ and $\overline{A_i} \cdot \overline{B_i} \cdot A_i \cdot B_i = 0$, from which

$$F_i = A_i \oplus B_i \oplus \overline{C_i} = \overline{A_i \oplus B_i \oplus C_i}.$$

This is the complement of the result obtained in problem 5.E.1.
In the case of subtraction:

$$F_i = \overline{A_i + \overline{B}_i} \oplus \overline{A_i \cdot \overline{B}_i} \oplus \overline{C_i}$$
$$= [(A_i + \overline{B_i}) \cdot (\overline{A_i} + B_i) + \overline{A_i} \cdot B_i \cdot A_i \cdot \overline{B}_i] \oplus \overline{C_i}$$
$$= \overline{A_i \oplus B_i} \oplus \overline{C_i} = A_i \oplus B_i \oplus C_i,$$

which is the classic result of this operation.

6. For $M = 1$, we have $\overline{C_i} = 1$; consequently the carry does not arise again. We obtain $F_i = G_i \oplus P_i \oplus 1 = \overline{G_i \oplus P_i} = \overline{\overline{G_i} \oplus \overline{P_i}}$, hence:

$$F_i = \overline{(S_3 \cdot A_i \cdot B_i + S_2 \cdot A_i \cdot \overline{B_i}) \oplus (S_1 \cdot \overline{B_i} + S_0 \cdot B_i + A_i)}.$$

7. The table is given in Figure 5.34. We notice that in this case, the 74181 is equivalent to 4 *logical* operators with one or two inputs.

$S_3\, S_2\, S_1\, S_0$	F_i	functions	$S_3\, S_2\, S_1\, S_0$	F_i	functions
0 0 0 0	$\overline{A_i}$		1 0 0 0	$\overline{A_i} + B_i$	B_i implied by A_i
0 0 0 1	$\overline{A_i + B_i}$	NOR	1 0 0 1	$\overline{A_i \oplus B_i}$	coincidence
0 0 1 0	$\overline{A_i} \cdot B_i$	B_i inh. by A_i	1 0 1 0	B_i	
0 0 1 1	0		1 0 1 1	$A_i \cdot B_i$	AND
0 1 0 0	$\overline{A_i \cdot B_i}$	NAND	1 1 0 0	1	
0 1 0 1	$\overline{B_i}$		1 1 0 1	$A_i + \overline{B_i}$	A_i implied by B_i
0 1 1 0	$A_i \oplus B_i$	Exclusive OR	1 1 1 0	$A_i + B_i$	OR
0 1 1 1	$A_i \cdot \overline{B}_i$	A_i inh. by B_i	1 1 1 1	A_i	

Figure 5.34

5.S.2.2 Arithmetic aspect

Addition

8a. We know that $NA = a_3 2^3 + a_2 2^2 + a_1 2^1 + a_0 2^0$ from which:

$$N\bar{A} = \bar{a}_3 2^3 + \bar{a}_2 2^2 + \bar{a}_1 2^1 + \bar{a}_0 2^0$$
$$= (1 - a_3)2^3 + (1 - a_2)2^2 + (1 - a_1)2^1 + (1 - a_0)2^0$$
$$= 2^3 + 2^2 + 2^1 + 2^0 - (a_3 2^3 + a_2 2^2 + a_1 2^1 + a_0 2^0), \quad \text{that is:}$$
$$N\bar{A} = 2^4 - 1 - NA.$$

Therefore the 74181 realizes:

$$C_{N+4} 2^4 + (2^4 - 1 - NF) = (2^4 - 1 - NA) + (2^4 - 1 - NB) + C_N, \text{ hence:}$$
$$NF = (C_{N+4} - 1)2^4 + N_A + N_B + (1 - C_N).$$
$$= (C_{N+4} - 1)2^4 + N_A + N_B + \overline{C_N}. \qquad (1)$$

Comment:

(a) NF represents the interpretation of the output word F in unsigned binary code.

(b) This relation represents the operation of the component for addition.

8b. Relation (1) becomes:

$$NS = (C_{N+4} - 1)2^4 + NX + NY + \overline{C_n}.$$

Putting $\overline{C_N} = C_0$:

— If $C_{N+4} = 1$, $NS = NX + NY + C_0$; the result does not overflow.

— If $C_{N+4} = 0$, $NS = -2^4 + NX + NY + C_0$; the result overflows.

C_{N+4} is therefore a possible carry to be propagated. Putting $C_4 = \overline{C_{N+4}}$, we thus obtain a classical adder (Figure 5.35).

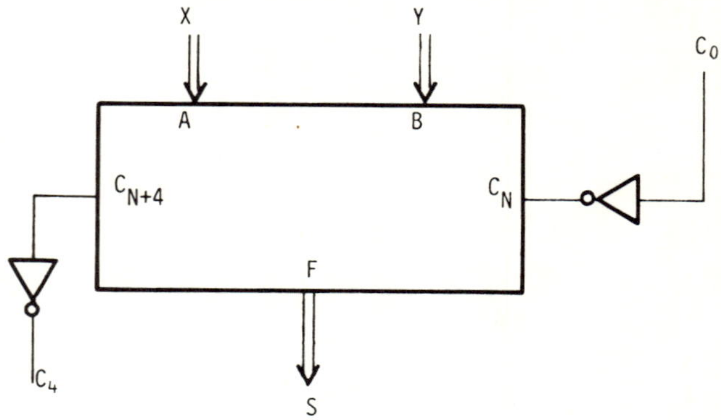

Figure 5.35

Example 1: $X = 0110$ and $Y = 0111$.

We have $NX = 6$ and $NY = 7$, hence $NS = 6 + 7 + C_0 = 13 + C_0$.

$$\text{If}\begin{cases} C_0 = 0, & NS = 13 \\ C_0 = 1, & NS = 14. \end{cases}$$

The 74181 realizes:

$$\begin{array}{llll} C_0 & = 0 \rightarrow C_N & = 1 \rightarrow & 1 \\ X & = 0110 & \rightarrow & 1001 \\ Y & = 0111 & \rightarrow & 1000 \end{array}$$

$$\begin{array}{lll} 1\ \ 0010 & \rightarrow & S = 1101 \\ \hookrightarrow C_{N+4} = 1 & \rightarrow C_4 = 0 & \end{array}$$

The result is indeed 13, without overflow.

$$\begin{array}{lll} C_0 = 1 \rightarrow C_N & 0 \rightarrow & 0 \\ X\ = 0110 & \rightarrow & 1001 \\ Y\ = 0111 & \rightarrow & 1000 \end{array}$$

$$\begin{array}{lll} 1\ \ 0001 & \rightarrow & S = 1110 \\ \hookrightarrow C_{N+4} & \rightarrow C_4 = 0 & \end{array}$$

The result is 14, without overflow.

Example 2: $X = 1001$ and $Y = 1000$.

We have $NX = 9$ and $NY = 8$, hence $NS = 9 + 8 + C_0 = 17 + C_0$.

$$\text{If}\begin{cases} C_0 = 0, & NS = 17 = 16 + 1 \\ C_0 = 1, & NS = 18 = 16 + 2. \end{cases}$$

The 74181 realizes:

$$
\begin{array}{lll}
C_0 = 0 \rightarrow C_N = 1 \rightarrow & 1 & \\
X = 1001 & \rightarrow 0110 & \\
Y = 1000 & \rightarrow 0111 & \\
\hline
 & 0\ \ 1110 & \rightarrow\ S = 0001 \\
 & \hookrightarrow C_{N+4} = 0 & \rightarrow C_4 = 1
\end{array}
$$

The result is 1 modulo 16, $C_4 = 1$ of weight 16, indicating an overflow.

$$
\begin{array}{lll}
C_0 = 1 \rightarrow C_N = 0 \rightarrow 0 & & \\
X = 1001 & \rightarrow 0110 & \\
Y = 1000 & \rightarrow 0111 & \\
\hline
 & 0\ \ 1101 & \rightarrow\ S = 0010 \\
 & \hookrightarrow C_{N+4} & \rightarrow C_4 = 1
\end{array}
$$

The result is 2 modulo 16, $C_4 = 1$ of weight 16, indicating an overflow.
8c. Relation (1) becomes:

$$NS = (C_{N+4} - 1)2^4 + (2^4 - 1 - NX) + (2^4 - 1 - NY) + \overline{C_N},$$

where, by replacing $\overline{C_N}$ by $1 - C_N$:

$$N \ = C_{N+4}2^4 - (NX + NY + C_N) + 2^4 - 1,$$

or $2^4 - 1 - NS = -2^4 C_{N+4} + NX + NY + C_N$.

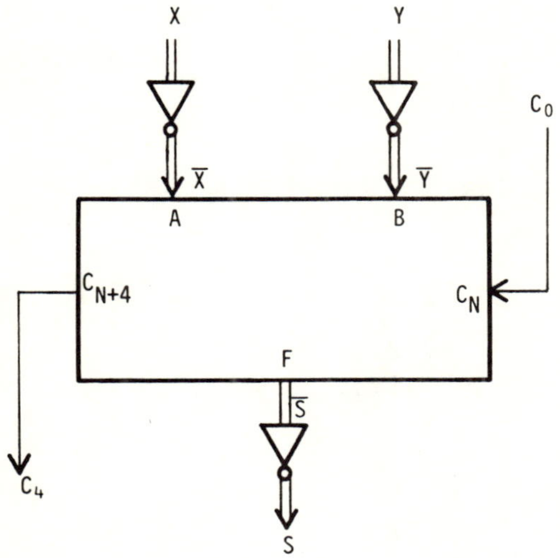

Figure 5.36

Now $2^4 - 1 - NS = N\bar{S}$, so we have:

$$N\bar{S} = -2^4 C_{N+4} + NX + NY + C_N.$$

Putting $C_N = C_0$:

— If $C_{N+4} = 0$, $N\bar{S} = NX + NY + C_0$; the result does not overflow and appears complemented.

— If $C_{N+4} = 1$, $N\bar{S} = -2^4 + NX + NY + C_0$; the result overflows.

C_{N+4} is therefore a possible carry to be propagated. Putting $C_4 = C_{N+4}$, we thus obtain a classical adder (Figure 5.36).

Reconsider the preceding examples with a different interpretation.

Example 1: $\bar{X} = 1001$ and $\bar{Y} = 1000$.

We have $NX = 6$ and $NY = 7$, so $NS = 6 + 7 + C_0 = 13 + C_0$.

We have seen that with these two words, the results given by the 74181 are:

$$\text{—If } C_0 = C_N = 0, \begin{cases} \bar{S} = 0010 \rightarrow NS = 13, \\ C_{N+4} = C_4 = 0 \rightarrow \text{no overflow} \end{cases}$$

$$\text{—If } C_0 = C_N = 1, \begin{cases} \bar{S} = 0001 \rightarrow NS = 14, \\ C_{N+4} = C_4 = 0 \rightarrow \text{no overflow} \end{cases}$$

Example 2: $\bar{X} = 0110$ and $\bar{Y} = 0111$.

We have $NX = 9$ and $NY = 8$, so $NS = 9 + 8 + C_0$.

We have seen that with these two words, the results given by the 74181 are:

$$\text{—If } C_0 = C_N = 0, \begin{cases} \bar{S} = 1110 \rightarrow NS = 1 \text{ modulo } 16, \\ C_{N+4} = C_4 = 1 \rightarrow \text{overflow} \end{cases}$$

$$\text{—If } C_0 = C_N = 1, \begin{cases} \bar{S} = 1101 \rightarrow NS = 2 \text{ modulo } 16, \\ C_{N+4} = C_4 = 1 \rightarrow \text{overflow} \end{cases}$$

Subtraction

9a. The 74181 realizes

$$+C_{N+4}2^4 + 2^4 - 1 - NF = (2^4 - 1 - NA) + [2^4 - 1 - (2^4 - 1 - NB)] + C_N$$

The changes of sign are needed for the replacement of the subtraction by an addition.

From which: $$NF = C_{N+4}2^4 + NA - NB - C_N \quad (2)$$

Comment:

(a) NF represents the interpretation of the output word F, in unsigned binary code.

(b) This relation represents the operation of the component for subtraction.

9b. *X and Y active high*

Calling D the difference word, B_4 the carry output, and B_0 the carry input, relation (2) becomes:

$$ND = B_4 2^4 + NX - NY - B_0.$$

— If $C_{N+4} = 0$, the result is positive.
— If $C_{N+4} = 1$, the result is negative; it can be written:

$$-2^4 + ND = NX - NY - B_0.$$

Now $-2^4 + ND = ND^*$, D represented in 2's complement. The result, therefore, can be interpreted.

Putting $B_4 = C_{N+4}$, we thus obtain a classical subtractor (Figure 5.37)

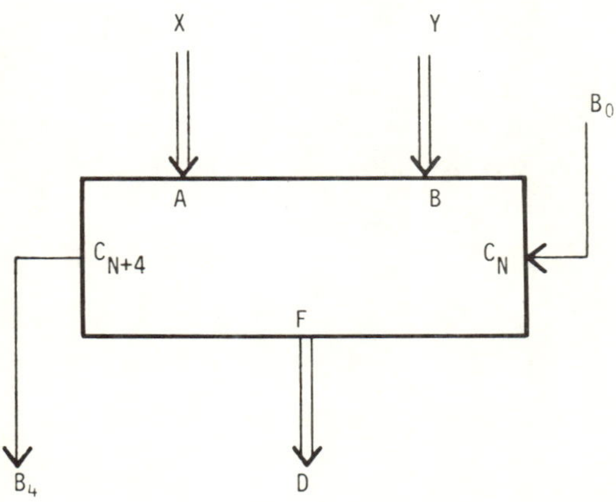

Figure 5.37

Example 1: $X = 1100$ and $Y = 0111$.
We have $NX = 12$ and $NY = 7$, so $ND = 12 - 7 - B_0 = 5 - B_0$.

$$\text{If } \begin{cases} B_0 = 0, & ND = 5 \\ B_0 = 1, & ND = 4. \end{cases}$$

The 74181 realizes:

$$B_0 = 0 \rightarrow C_N = 0 \rightarrow 0$$
$$X = 1100 \quad \rightarrow \quad 0011$$
$$Y = 0111 \quad \rightarrow \quad 0111$$

$$0\ 1010 \qquad \rightarrow D = 0101$$
$$\rightarrow C_{N+4} = 0 \quad \rightarrow B_4 = 0$$

The result is 5 and positive.
With $B_0 = 1$, it is easy to obtain: $D = 0100$, $B_4 = 0$, hence the result is 4.

Example 2: $X = 0111$ and $Y = 1100$.
We have $NX = 7$ and $NY = 12$, so $ND = 7 - 12 - B_0 = -5 - B_0$.

$$\text{If } \begin{cases} B_0 = 0, & ND = -5 \\ B_0 = 1, & ND = -6 \end{cases}$$

The 74181 realizes:

$$\begin{array}{lll} B_0 = 0 \rightarrow C_N = 0 \rightarrow 0 & & \\ X = 0111 & \rightarrow 1000 & \\ Y = 1100 & \rightarrow 1100 & \\ \hline & 1\ 0100 & \rightarrow D = 1011 \\ & \hookrightarrow C_{N+4} = 1 \rightarrow B_4 = 1 & \end{array}$$

The result is negative; in 2's complement we have $ND^* = -5$. With $B_0 = 1$, it is easy to obtain: $D = 1010$, $B_4 = 1$, hence the result is -6.

9b. (contd.) *X and Y active low*
Relation (2) becomes:

$$\begin{aligned} ND &= C_{N+4}2^4 + (2^4 - 1 - NX) - (2^4 - 1 - NY) - CN \\ &= C_{N+4}2^4 - NX + NY - C_N. \end{aligned}$$

— If $C_{N+4} = 1$, $ND = 2^4 - NX + NY - C_N$, hence:

$$2^4 - 1 - ND = NX - NY - (1 - C_N),$$

putting: $1 - C_N = \overline{C_N} = B_0$:

$$N\bar{D} = NX - NY - B_0.$$

Thus the result D is positive, but appears in complemented form $\bar{D}$ at the output.

$$\text{—If } C_{N+4} = 0, \; ND = -NX + NY - C_N = NY - NX - C_N.$$

The result is negative, but output D provides the modulus.
Putting $B_4 = \overline{C_{N+4}}$, we thus obtain a classical subtractor (Figure 5.38).

Example 1: $\bar{X} = 0111$ and $\bar{Y} = 1100$.
We have $NX = 8$ and $NY = 3$, so $ND = 8 - 3 - B_0 = 5 - B_0$.
We have seen that with these two words, the results given by the 74181 are:

$$\text{—If } B_0 = \overline{C_N} = 0, \text{ then } C_N = 1, \begin{cases} \bar{D} = 1010 \rightarrow ND = 5 \\ C_{N+4} = 1 \rightarrow B_4 = 0 \end{cases}$$

$$\text{—If } B_0 = \overline{C_N} = 1, \text{ then } C_N = 0, \begin{cases} \bar{D} = 1011 \rightarrow ND = 4 \\ C_{N+4} = 1 \rightarrow B_4 = 0. \end{cases}$$

In both cases the result is positive.

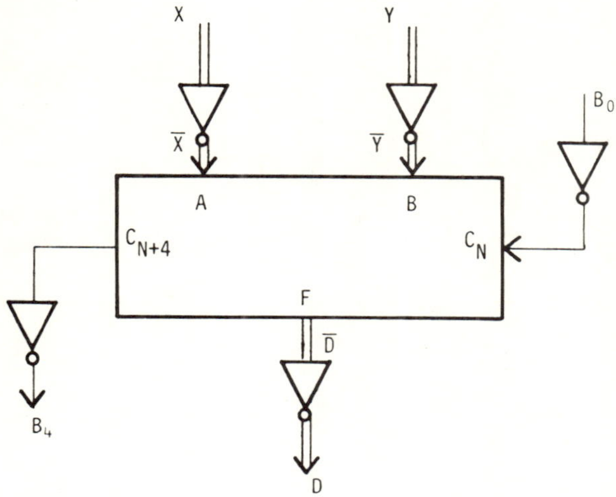

Figure 5.38

Example 2: $\bar{X} = 1100$ and $\bar{Y} = 0111$.
We have $NX = 3$ and $NY = 8$, so $ND = 3 - 8 - B_0 = -5 - B_0$.

We have seen that with these two words, the results given by the 74181 are

$$\text{— If } B_0 = \overline{C_N} = 0, \text{ then } C_N = 1, \begin{cases} \bar{D} = 0100 \rightarrow N\bar{D} = 4 \\ C_{N+4} = 0 \rightarrow B_4 = 1 \end{cases}$$

$$\text{— If } B_0 = \overline{C_N} = 1, \text{ then } C_N = 0, \begin{cases} \bar{D} = 0101 \rightarrow N\bar{D} = 5 \\ C_{N+4} + 0 \rightarrow B_4 = 1 \end{cases}$$

In both cases the result is negative.

9c. Figures 5.35, 5.36, 5.37, and 5.38 summarize the operation.

Connection of several components

10. Combining these figures and seeking the simplest solution, we obtain Figure 5.39.

Switches at H: The data are active high.
— If OP = 1, the operation realized is addition. We see that the carry input is 1 and that the carry output is complemented. This is modelled by a cascade connection of four arrangements identical to that of Figure 5.35, after simplification of the two inverters in series between each 74181.
— If OP = 0, the operation realized is subtraction. We see that the carry input is 0. The carry output is not complemented. This is modelled by a cascade connection of four arrangements identical to that of Figure 5.37.

Switches at L: The data are active low.
— If OP = 1, the operation realized is addition. We see that the carry input is

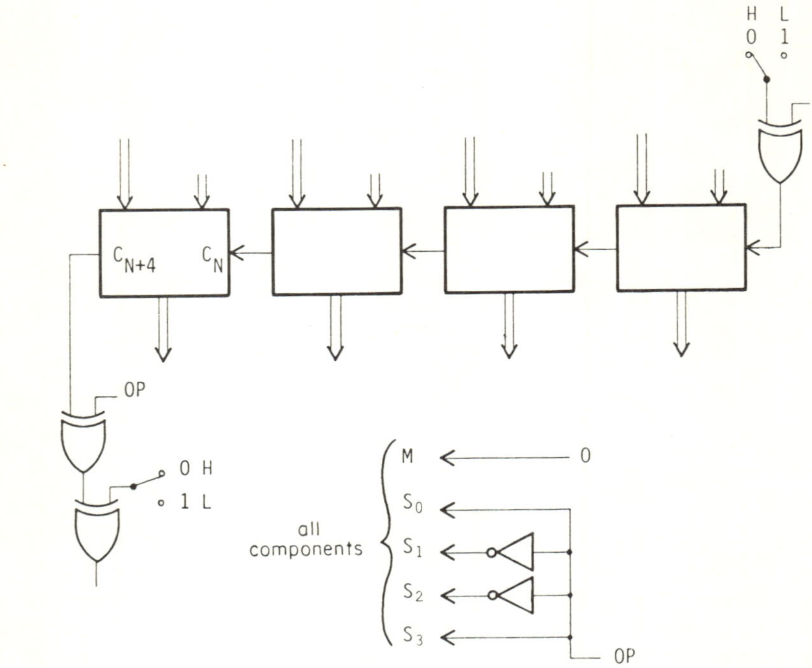

Figure 5.39

0 and that the carry output is not complemented. This is modelled by a cascade connection of four arrangements identical to that of Figure 5.36.

— If $OP = 0$, the operation realized is subtraction. We see that the carry input is 1 and that the carry output is complemented. This is modelled by a cascade connection of four arrangements identical to that of Figure 5.38, after simplification of the two inverters in series between each 74181.

5.S.3 Operation of the carry look-ahead circuit 74182 or 14582B

1. We obtain:

$$\bar{P} = \overline{P_3 \cdot P_2 \cdot P_1 \cdot P_0} \qquad \text{and} \qquad \bar{G} = \overline{G_3 + P_3 \cdot G_2 + P_3 \cdot P_2 \cdot G_1 + P_3 \cdot P_2 \cdot P_1 \cdot G_0}$$

$$C_{N+4} = G + P_3 \cdot P_2 \cdot P_1 \cdot P_0 \cdot C_N$$

$$= G_3 + P_3 \cdot G_2 + P_3 \cdot P_2 \cdot G_1 + P_3 \cdot P_2 \cdot P_1 \cdot G_0 + P_3 \cdot P_2 \cdot P_1 \cdot P_0 \cdot C_N,$$

a relation demonstrated in problem 5.E.1.

2. We obtain successively:

$$C_{N+X} = \overline{\overline{G_0} \cdot \overline{P_0} + \overline{G_0} \cdot \overline{C_N}} = \overline{\overline{G_0} \cdot (\overline{P_0} + \overline{C_N})} = G_0 + P_0 \cdot C_N$$

$$C_{N+Y} = \overline{\overline{G_1} \cdot \overline{P_1} + \overline{G_1} \cdot \overline{G_0} \cdot \overline{P_0} + \overline{G_1} \cdot \overline{G_0} \cdot \overline{C_N}}$$

$$= \overline{\overline{G_1}\cdot[\overline{P_1} + \overline{G_0}\cdot(\overline{P_0} + \overline{C_N})]}$$
$$= G_1 + P_1\cdot(G_0 + P_0\cdot C_N) = G_1 + P_1\cdot G_0 + P_1\cdot P_0\cdot C_N$$
$$C_{N+Z} = \overline{\overline{G_2}\cdot\overline{P_2} + \overline{G_2}\cdot\overline{G_1}\cdot\overline{P_1} + \overline{G_2}\cdot\overline{G_1}\cdot\overline{G_0}\cdot\overline{P_0} + \overline{G_2}\cdot\overline{G_1}\cdot\overline{G_0}\cdot\overline{C_N}}$$
$$= \overline{\overline{G_2}\cdot[\overline{P_2} + \overline{G_1}\cdot[\overline{P_1} + \overline{G_0}\cdot(\overline{P_0} + \overline{C_N})]]}$$
$$= G_2 + P_2\cdot[G_1 + P_1\cdot(G_0 + P_0\cdot C_N)]$$
$$= G_2 + P_2\cdot G_1 + P_2\cdot P_1\cdot G_0 + P_2\cdot P_1\cdot P_0\cdot C_N$$
$$\overline{P} = \overline{P_3} + \overline{P_2} + \overline{P_1} + \overline{P_0} = \overline{P_3\cdot P_2\cdot P_1\cdot P_0}$$
$$\overline{G} = \overline{G_3}\cdot\overline{P_3} + \overline{G_3}\cdot\overline{G_2}\cdot\overline{P_2} + \overline{G_3}\cdot\overline{G_2}\cdot\overline{G_1}\cdot\overline{P_1} + \overline{G_3}\cdot\overline{G_2}\cdot\overline{G_1}\cdot\overline{G_0}$$
$$= \overline{G_3}\cdot[\overline{P_3} + \overline{G_2}\cdot[\overline{P_2} + \overline{G_1}\cdot(\overline{P_1} + \overline{G_0})]]$$

hence, considering C_{N+Z}:

$$\overline{G} = \overline{G_3 + P_3\cdot G_2 + P_3\cdot P_2\cdot G_1 + P_3\cdot P_2\cdot P_1\cdot G_0}$$

3. Figure 5.40 shows that the maximum calculation time is 107 ns, taking the worst case. This time is obtained in the following way: the 74181–0 establishes R_4 after a maximum of 50 ns since both 16 bits of data are applied simultaneously. After that only the time $C_N \Rightarrow C_{N+4}$ intervenes. We can consider that the operation is finished when R_{16} is obtained, that is after $50 + (19 \times 3) = 107$ ns.

4. In order to avoid confusion, we shall indicate the inputs of the 74182 by G' and P'.

4a. We have $C_{N+X} = G'_0 + P'_0\cdot C_N$. Now:

$$74181-0 \begin{cases} G'_0 = G_3 + P_3\cdot G_2 + P_3\cdot P_2\cdot G_1 + P_3\cdot P_2\cdot P_1\cdot G_0 \\ P'_0 = P_3\cdot P_2\cdot P_1\cdot P_0 \\ C_N = C_0 \end{cases}$$

Substituting:

$$C_{N+X} = G_3 + P_3\cdot G_2 + P_3\cdot P_2\cdot G_1 + P_3\cdot P_2\cdot P_1\cdot G_0 + P_3\cdot P_2\cdot P_1\cdot P_0\cdot C_0,$$

which is the relation giving R_4.

Similarly: $C_{N+Y} = G'_1 + P'_1\cdot G'_0 + P'_1\cdot P'_0\cdot C_N$ Now:

$$74181-1 \begin{cases} G'_1 = G_7 + P_7\cdot G_6 + P_7\cdot P_6\cdot G_5 + P_7\cdot P_6\cdot P_5\cdot G_4 \\ P'_1 = P_7\cdot P_6\cdot P_5\cdot P_4 \\ C_N = C_0 \end{cases}$$

Substituting:

$$\begin{aligned} C_{N+Y} = {} & G_7 + P_7\cdot G_6 + P_7\cdot P_6\cdot G_5 + P_7\cdot P_6\cdot P_5\cdot G_4 + P_7\cdot P_6\cdot P_5\cdot P_4\cdot G_3 \\ & + P_7\cdot P_6\cdot P_5\cdot P_4\cdot P_3\cdot G_2 + P_7\cdot P_6\cdot P_5\cdot P_4\cdot P_3\cdot P_2\cdot G_1 \\ & + P_7\cdot P_6\cdot P_5 P_6\cdot P_4\cdot P_3\cdot P_2\cdot P_1\cdot G_0 \\ & + P_7\cdot P_6\cdot P_5\cdot P_4\cdot P_3\cdot P_2\cdot P_1\cdot P_0\cdot C_0, \end{aligned}$$

which is the relation giving R_8.

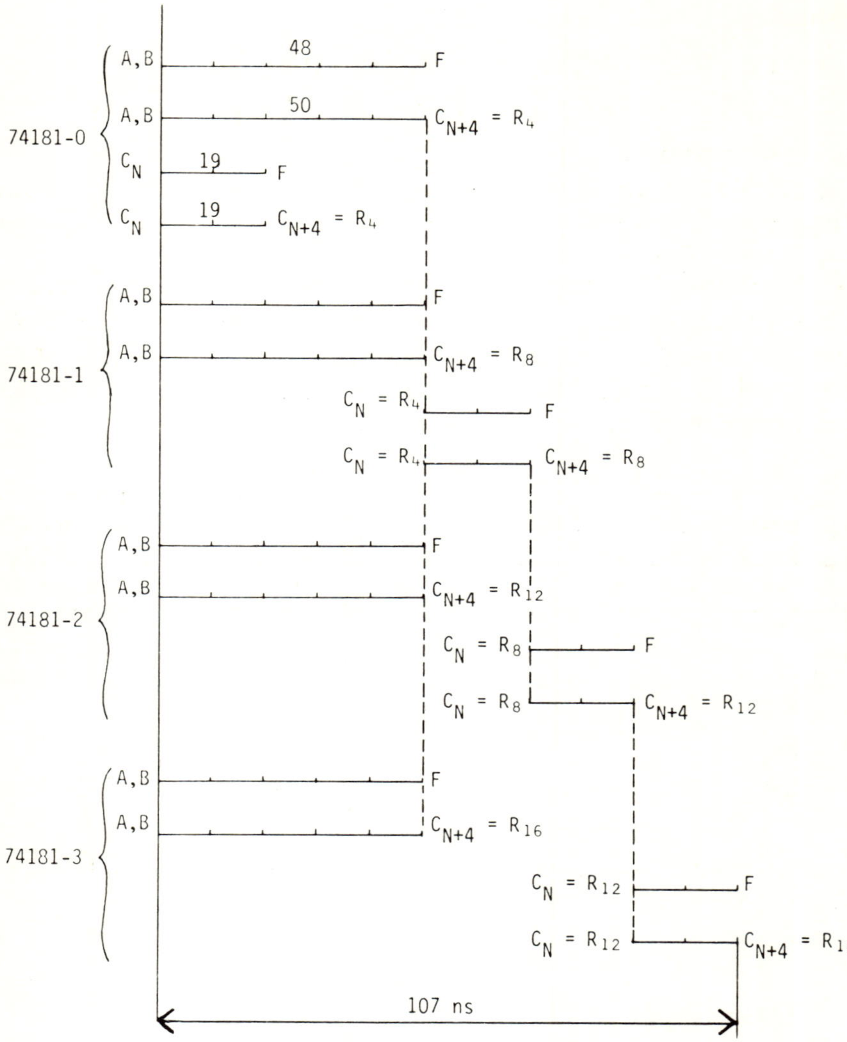

Figure 5.40

Similarly:

$$C_{N+Z} = G'_2 + P'_2 \cdot G'_1 + P'_2 \cdot P'_1 \cdot G'_0 + P'_2 \cdot P'_1 \cdot P'_0 \cdot C_N.$$

Now:

$$74181-2 \begin{cases} G'_2 = G_{11} + P_{11} \cdot G_{10} + P_{11} \cdot P_{10} \cdot G_9 + P_{11} \cdot P_{10} \cdot P_9 \cdot G_8 \\ P'_2 = P'_{11} \cdot P'_{10} \cdot P'_9 \cdot P'_8 \\ C_N = C_0 \end{cases}$$

Substituting:

$$C_{N+Z} = G_{11} + P_{11}\cdot G_{10} + P_{11}\cdot P_{10}\cdot G_9 + P_{11}\cdot P_{10}\cdot P_9\cdot G_8 + \ldots$$
$$+ P_{11}\cdot P_{10}\cdot P_9\cdot P_8\cdot P_7\cdot P_6\cdot P_5\cdot P_4\cdot P_3\cdot P_2\cdot P_1\cdot P_0\cdot C_0$$

which is the relation giving R_{12}.

4b. For a 16-bit format, C_{N+Z} does not depend on $\overline{G_3}$ or $\overline{P_3}$. These two outputs are necessary only to obtain $\bar{G}$ and $\bar{P}$ in order to extend to a format greater than 16 bits.

4c. The diagram is given in Figure 5.41. It consists of three arrangements of the type given in Figure 5.12, but with $\bar{G}$ and $\bar{P}$ of 74181–3 connected to $\overline{G_3}$ and $\overline{P_3}$ of the 74182. A fourth 74182 receives the $\bar{G}$ and $\bar{P}$ outputs of the other 74182's, in order to obtain a look-ahead carry for the 48 bits. Only the necessary inputs and outputs have been used.

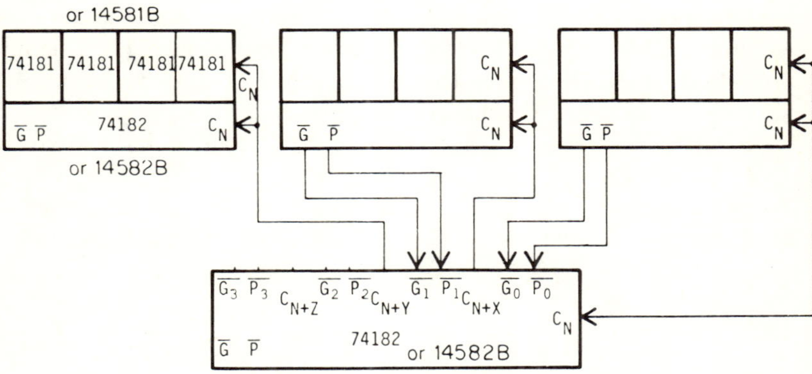

Figure 5.41 The three arrangements above are identical to that of Figure 5.12, but with $\bar{G}$ and $\bar{P}$ of 74181-3 connected to $\bar{G}_3$ and $\bar{P}_3$ of the 74182.

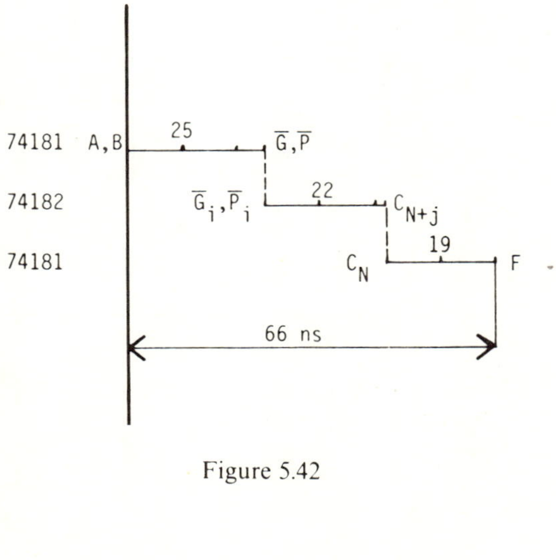

Figure 5.42

4d. Since 5.42 shows that the maximum calculation time is 66 ns, considering the worst case. In this hypothesis, all the 74181's operate simultaneously without waiting for the carry inputs C_N provided by the preceding 74181.

4e. For 16 bits, the difference is already appreciable; the reduction in calculation time is:

$$100 \times \frac{(107 - 66)}{107} \simeq 38\% \text{ less}$$

5.S.4 Decimal arithmetic in BCD code

5.S.4.1 Study of addition

1. Zone IV does not arise because the largest value obtained is $9 + 9 + 1 = 19_{10} = 13_{16}$. 1 represents the carry provided by a preceding stage.

2. The difference between the *actual working base* of the adder 16 and the *conventional working base* 10 is 6; it is sufficient, therefore, to add 6 to the values in zone II of the table of Figure 5.15(a) to recover those of zone II of the table of Figure 5.14(a). This addition systematically creates a carry which must be retained in order to recover Figure 5.14(b).

3. The correction is the same, however no carry is created, but the existing one (Figure 5.15(b)) is to be retained.

4. It is sufficient to notice that C_{n+1} appears (zone II) or exists (zone III) when the addition of 6 is necessary, hence the diagram of Figure 5.43 and the K map of Figure 5.44.

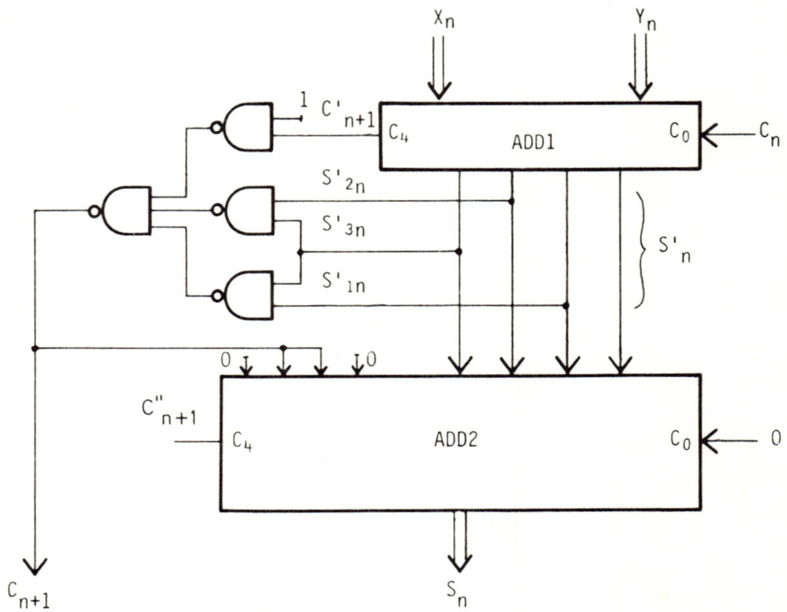

Figure 5.43

We notice that $C_{n+1} = C'_{n+1} + C''_{n+1}$.

$S'_{1n}S'_{0n}$ \ $C'_{n+1}S'_{3n}S'_{2n}$	000	001	011	010	110	111	101	100
00	0	0	1	0	-	-	-	1
01	0	0	1	0	-	-	-	1
11	0	0	1	1	-	-	-	1
10	0	0	1	1	-	-	-	1

Figure 5.44 Command to add $+6 = C_{n+1} = C'_{n+1} + S'_{3n}\cdot S'_{2n} + S'_{3n}\cdot S'_{1n}$

5. It is sufficient to add $+6$ permanently and to multiplex the outputs of the two adders (Figure 5.45).

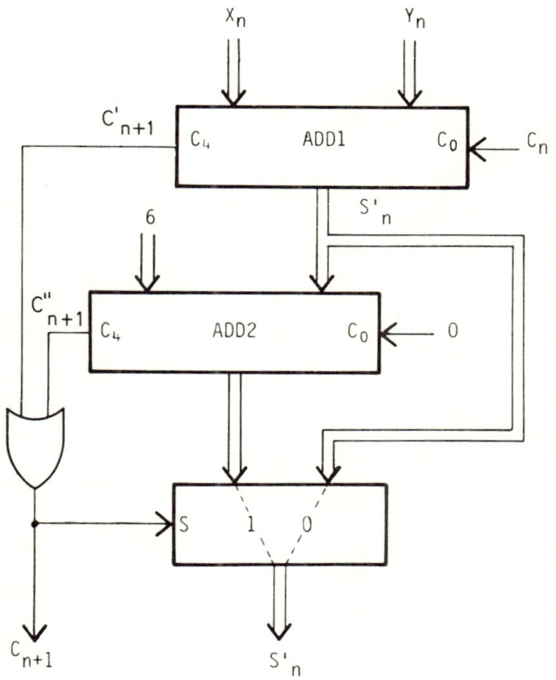

Figure 5.45

6. From the tables of Figures 5.14 and 5.15, we can set up the truth table of Figure 5.46, from which the following expressions can be deduced.

$$C_{n+1} = C'_{n+1} + S'_{3n}\cdot(S'_{2n} + S'_{1n}) \qquad \text{(From Figure 5.44)}$$

$$S_{3n} = S'_{3n}\cdot\bar{S}'_{2n}\cdot\bar{S}'_{1n} + C'_{n+1}\cdot S'_{1n},$$

$$S_{2n} = \bar{S}'_{3n}\cdot S'_{2n} + S'_{2n}\cdot S'_{1n} + \bar{S}'_{1n}\cdot C'_{n+1},$$

$$S_{1n} = S'_{3n}\cdot S'_{2n}\cdot\bar{S}'_{1n} + \bar{S}'_{3n}\cdot S'_{1n}\cdot\bar{C}'_{n+1} + \bar{S}'_{1n}\cdot C'_{n+1},$$

$$S_{0n} = S'_{0n}.$$

16	8	4	2	1	10	8	4	2	1	
C'_{n+1}	S'_{3n}	S'_{2n}	S'_{1n}	S'_{0n}	C_{n+1}	S_{3n}	S_{2n}	S_{1n}	S_{0n}	
0	0	0	0	0	0	0	0	0	0	
0	0	0	0	1	0	0	0	0	1	
0	0	0	1	0	0	0	0	1	0	
0	0	0	1	1	0	0	0	1	1	
0	0	1	0	0	0	0	1	0	0	Zone I
0	0	1	0	1	0	0	1	0	1	
0	0	1	1	0	0	0	1	1	0	
0	0	1	1	1	0	0	1	1	1	
0	1	0	0	0	0	1	0	0	0	
0	1	0	0	1	0	1	0	0	1	
0	1	0	1	0	1	0	0	0	0	
0	1	0	1	1	1	0	0	0	1	
0	1	1	0	0	1	0	0	1	0	Zone II
0	1	1	0	1	1	0	0	1	1	
0	1	1	1	0	1	0	1	0	0	
0	1	1	1	1	1	0	1	0	1	
1	0	0	0	0	1	0	1	1	0	
1	0	0	0	1	1	0	1	1	1	Zone III
1	0	0	1	0	1	1	0	0	0	
1	0	0	1	1	1	1	0	0	1	
1	0	1	0	0	—	—	—	—	—	
1	0	1	0	1	—	—	—	—	—	
1	0	1	1	0	—	—	—	—	—	
1	0	1	1	1	—	—	—	—	—	
1	1	0	0	0	—	—	—	—	—	
1	1	0	0	1	—	—	—	—	—	Zone IV
1	1	0	1	0	—	—	—	—	—	
1	1	0	1	1	—	—	—	—	—	
1	1	1	0	0	—	—	—	—	—	
1	1	1	0	1	—	—	—	—	—	
1	1	1	1	0	—	—	—	—	—	
1	1	1	1	1	—	—	—	—	—	

Figure 5.46

Figure 5.47 shows a BCD circuit in CMOS technology.

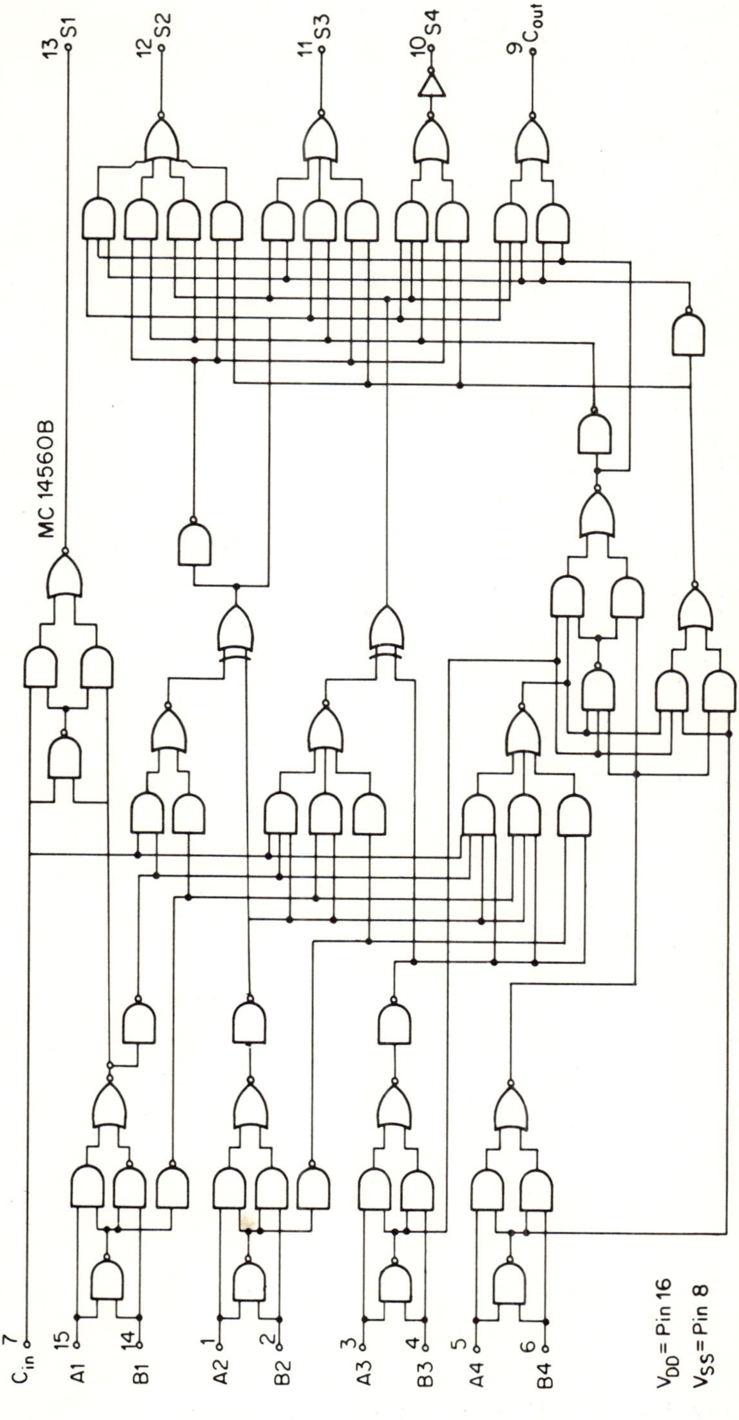
MC 14560B
7 C_{in}
15 A1
14 B1
1 A2
2 B2
3 A3
4 B3
5 A4
6 B4
13 S1
12 S2
11 S3
10 S4
9 C_{out}
V_{DD} = Pin 16
V_{SS} = Pin 8

Figure 5.47

TRUTH TABLE*

INPUT									OUTPUT				
A4	A3	A2	A1	B4	B3	B2	B1	C_{in}	C_{out}	S4	S3	S2	S1
0	0	0	0	0	0	0	0	0	0	0	0	0	0
0	0	0	0	0	0	0	0	1	0	0	0	0	1
0	1	0	0	0	0	1	1	0	0	0	1	1	1
0	1	0	0	0	0	1	1	1	0	1	0	0	0
0	1	1	1	0	1	0	0	0	1	0	0	0	1
0	1	1	1	0	1	0	0	1	1	0	0	1	0
1	0	0	0	0	1	0	1	0	1	0	0	1	1
0	1	1	0	1	0	0	0	0	1	0	1	0	0
1	0	0	1	1	0	0	1	1	1	1	0	0	1

* Partial truth table to show logic operation for representative input values.

5.S.5 Addition and subtraction of numbers expressed as modulus and sign

5.S.5.1 Addition

1. Let X and Y be the models of the two numbers with:

$$X{:}s_X x_{n-1} x_{n-2} \ldots x_1 x_0$$

$$Y{:}s_y y_{n-1} y_{n-2} \ldots y_1 y_0$$

where s_X and s_Y are the signs.

Putting:

$$|X| = x_{n-1} \ldots x_1 = \text{modulus of } X, \quad |Y| = y_{n-1} \ldots y_1 = \text{modulus of } Y,$$

from which $NX = \sum_{i=0}^{n-1} x_i \times 2^i$ and $NY = \sum_{i=0}^{n-1} y_i \times 2^i$.

Putting also:

$$R{:}s_R r_{n-1} \ldots r_0, \quad \text{a model of the result,}$$

from which $|R| = r_{n-1} \ldots r_0 = $ modulus of R and $NR = \sum_{i=0}^{n-1} r_i \times 2^i$.

Let us analyse the operation. There are four cases according to the respective signs of X and Y, and six cases taking account of the respective values of NX and NY. The representative table is given in Figure 5.48.

sign comparison	modulus comparison	required operations on the modulus	required operations on the signs
1 $s_X = s_Y = 0$		$\|R\| = \|X\|$ plus $\|Y\|$	$s_R = s_X = s_Y = 0$
2 $s_X = s_Y = 1$		$\|R\| = \|X\|$ plus $\|Y\|$	$s_R = s_X = s_Y = 1$
3 $s_X = 0$	a $NX > NY$	$\|R\| = \|X\| - \|Y\|$	$s_R = s_X = 0$
$s_Y = 1$	b $NX < NY$	$\|R\| = \|Y\| - \|X\|$	$s_R = s_Y = 1$
4 $s_X = 1$	a $NX > NY$	$\|R\| = \|X\| - \|Y\|$	$s_R = s_X = 1$
$s_Y = 0$	b $NX < NY$	$\|R\| = \|Y\| - \|X\|$	$s_R = s_Y = 0$

Figure 5.48

2. We notice that the solution implies the use of a binary adder–subtractor and comparison of the modulus of X and Y, so that the subtractor always gives a positive result. It is possible to find a satisfactory solution using this information, but it is more interesting to construct an architecture which notably simplifies the solution (Figure 5.49).

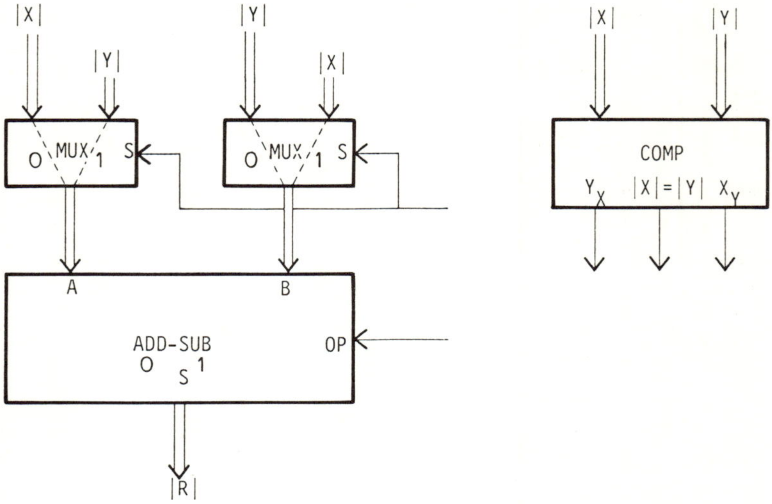

Figure 5.49

ADD–SUB is an adder–subtractor such that:

$$\text{if} \begin{cases} \text{OP} = 0, & \text{S} = A \text{ plus } B \\ \text{OP} = 1, & S = A - B \end{cases}$$

MUX is a multiplexer such that:

$$\text{if} \begin{cases} S = 0, & A = |X| \text{ and } B = |Y| \\ S = 1, & A = |Y| \text{ and } B = |X| \end{cases}$$

Thus we see that it is possible to realize

$$|R| = |X| - |Y| \quad \text{or} \quad |R| = |Y| - |X|$$

COMP is a comparator such that:

$$X_Y = 1 \text{ if } NX > NY$$
$$Y_X = 1 \text{ if } NX < NY.$$

3. From the architecture of Figure 5.49 and the table of Figure 5.48, we obtain the truth table of Figure 5.50.

We notice that four combinations are impossible: these arise when X_Y and Y_X are simultaneously at 1, since these variables are exclusive. On the other

s_X	s_Y	X_Y	Y_Y	case	comment		S	OP	s_R
0	0	0	0	} 1	$\|X\| = Y\|$		–	0	0
0	0	0	1				–	0	0
0	0	1	0				–	0	0
0	0	1	1	imp^{ble}			–	–	–
0	1	0	0	3b	$\|X\| = \|Y\|$	$NR = 0$	–	1	0
0	1	0	1	3a			1	1	1
0	1	1	0				0	1	0
0	1	1	1	imp^{ble}			–	–	–
1	0	0	0	4b	$\|X\| = \|Y\|$	$NR = 0$	–	1	0
1	0	0	1	4a			1	1	0
1	0	1	0				0	1	1
1	0	1	1	imp^{ble}			–	–	–
1	1	0	0	} 2	$\|X\| = Y\|$		–	0	1
1	1	0	1				–	0	1
1	1	1	0				–	0	1
1	1	1	1	imp^{ble}			–	–	–

Figure 5.50

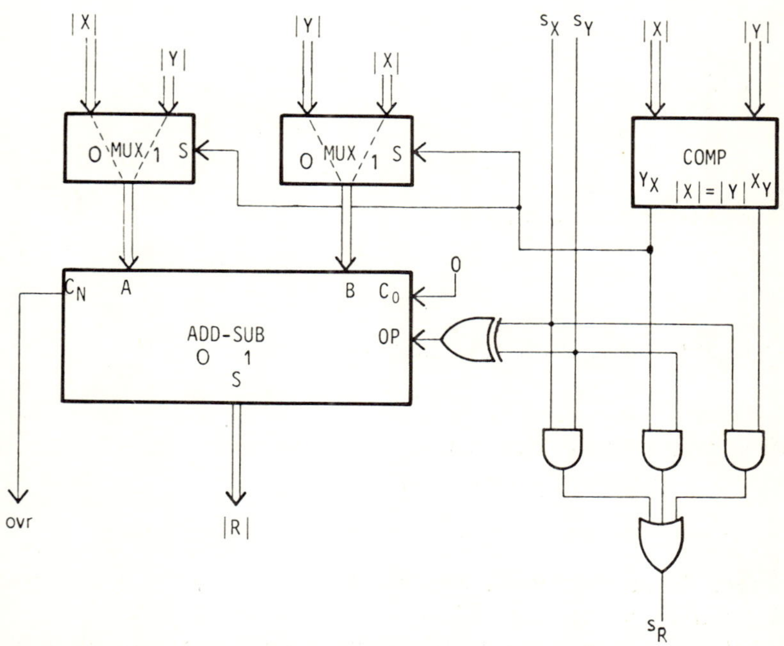

Figure 5.51

hand, when X_Y and Y_X are simultaneously at 0, that is $|X| = |Y|$ and therefore $|R| = 00\ldots0$, in the case of a subtraction; we have imposed $s_R = 0$, that is to say $(+0)\cdot(-0)$ no longer exists.

With the help of three K maps, which are not shown, we obtain:

$$s_R = s_X \cdot s_Y + s_X \cdot X_Y + s_Y \cdot Y_X, \qquad \mathrm{OP} = s_X \oplus s_Y, \qquad S = Y_X$$

The final diagram is given in Figure 5.51. The signal ovr indicates an overflow.

5.S.5.2 Subtraction

1. In the same way as before, we can set up the table of Figure 5.52.

sign comparison	modulus comparison	required operations on the modulus	required operations on the signs
1 $s_X = s_Y = 0$	a $NX > NY$	$\lvert R\rvert = \lvert X\rvert - \lvert Y\rvert$	$s_R = s_X = 0$
	b $NX < NY$	$\lvert R\rvert = \lvert Y\rvert - \lvert X\rvert$	$s_R = 1$
2 $s_X = s_Y = 1$	a $NX > NY$	$\lvert R\rvert = \lvert X\rvert - \lvert Y\rvert$	$s_R = s_X = 1$
	b $NX < NY$	$\lvert R\rvert = \lvert Y\rvert - \lvert X\rvert$	$s_R = 0$
3 $s_X = 0$, $s_Y = 1$		$\lvert R\rvert = \lvert X\rvert$ plus $\lvert Y\rvert$	$s_R = s_X = 0$
4 $s_X = 1$, $s_Y = 0$		$\lvert R\rvert = \lvert X\rvert$ plus $\lvert Y\rvert$	$s_R = s_X = 1$

Figure 5.52

2. We use the same architecture.
3. From the table of Figure 5.52, we obtain the truth table of Figure 5.53.

In the same way as previously, three K maps give:

$$s_R = s_X \cdot \overline{s_Y} + s_X \cdot X_Y + \overline{s_Y} \cdot Y_X, \qquad \mathrm{OP} = s_X \oplus \overline{s_Y}, \qquad S = Y_X$$

The diagram is easily deduced from these results.

Comment: It can be seen that in complementing s_Y we return to the previous case, without needing to repeat the whole study.

s_X	s_Y	X_Y	Y_X	case	comments		S	OP	s_R
0	0	0	0	1b	$\lvert X\rvert=\lvert Y\rvert$	$NR=0$	—	1	0
0	0	0	1	1a			1	1	1
0	0	1	0				0	1	0
0	0	1	1	imp.[ble]			—	—	—
0	1	0	0	}	$\lvert X\rvert=\lvert Y\rvert$		—	0	0
0	1	0	1	} 3			—	0	0
0	1	1	0	}			—	0	0
0	1	1	1	imp.[ble]			—	—	—
1	0	0	0		$\lvert X\rvert=\lvert Y\rvert$		—	0	1
1	0	0	1	4			—	0	1
1	0	1	0				—	0	1
1	0	1	1	imp.[ble]			—	—	—
1	1	0	0	2b	$\lvert X\rvert=\lvert Y\rvert$	$NR=0$	—	1	0
1	1	0	1	2a			1	1	0
1	1	1	0				1	1	0
1	1	1	1	imp.[ble]			—	—	—

Figure 5.53

5.S.5.3 Addition-subtraction

The comment above allows the final diagram of Figure 5.54 to be obtained immediately.

Collecting the results of the two studies:

$S/D=0$, ADD	$S/D=1$, SUB
$s_R = s_X\cdot s_Y + s_X\cdot X_Y + s_Y\cdot Y_X$	$s_R = s_X\cdot \bar{s}_Y + s_X\cdot X_Y + \bar{s}_Y\cdot Y_X$,
$\mathrm{OP} = s_X \oplus s_Y$	$\mathrm{OP} = s_X \oplus \bar{s}_Y$,
$S = Y_X$	$S = Y_X$

They can be regrouped in the form:

$$s_R = s_X\cdot X_Y + (s_Y \oplus S/D)\cdot s_X + (s_Y \oplus S/D)\cdot Y_X$$

$$\mathrm{OP} = s_x \oplus (s_Y \oplus S/D)$$

$$S = Y_X$$

$$\text{If} \qquad \begin{cases} S/D=0, \Rightarrow s_Y \oplus S/D = s_Y \\ S/D=1, \Rightarrow s_Y \oplus S/D = s_Y. \end{cases}$$

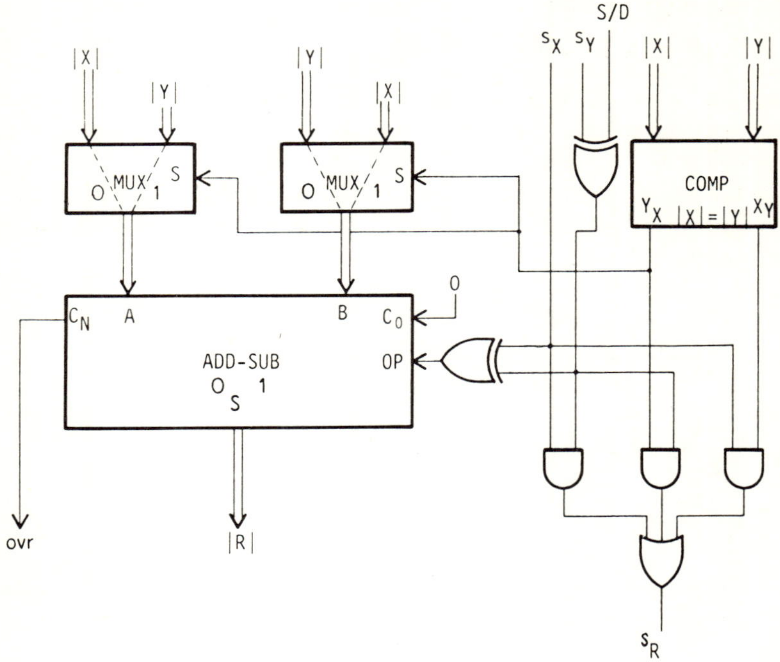

Figure 5.54

5.S.6 Addition and subtraction of numbers expressed in 2's complement form

Recall that signed and unsigned representations are associated with the model $X = x_{n-1}x_{n-2}\ldots x_0$; they are respectively:

$$NX = \sum_{i=0}^{n-1} x_i 2^i, \quad \text{and}$$

$$NX^* = -x_{n-1}2^{n-1} + \sum_{i=0}^{n-2} x_i 2^i.$$

Hence the relation $NX^* = -x_{n-1}2^n + NX$.

5.S.6.1 Addition

1. The use of a binary adder allows us to write:

$$C_n \times 2^n + NR = NX + NY + C_0, \tag{1}$$

where C_0 and C_n are the input and output carries.

Now:

$$NX = x_{n-1}2^n + NX^*,$$
$$NY = y_{n-1}2^n + NY^*,$$
$$NR = r_{n-1}2^n + NR^*.$$

Substituting in (1), we obtain:

$$NR^* = NX^* + NY^* + C_0 + (x_{n-1} + y_{n-1} - r_{n-1} - C_n)2^n.$$

Now rank n is situated beyond the sign, and therefore cannot exist. Consequently the algebraic quantity $(x_{n-1} + y_{n-1} - r_{n-1} - C_n)$ must be zero, to avoid overflow.

2. We know that the arithmetic sum modulo 2 and also the difference are logically equivalent to the exclusive OR operation; hence, it is possible to create an *overflow bit* such that:

$$\text{ovr} = x_{n-1} \oplus y_{n-1} \oplus r_{n-1} \oplus C_n.$$

It has been established, when studying binary addition, that:

$$r_{n-1} = x_{n-1} \oplus y_{n-1} \oplus C_{n-1},$$

hence: $\quad \text{ovr} = x_{n-1} \oplus y_{n-1} \oplus x_{n-1} \oplus y_{n-1} \oplus C_{n-1} \oplus C_n.$

Now: $\quad x_{n-1} \oplus y_{n-1} \oplus x_{n-1} \oplus y_{n-1} = 0,$

from which: $\quad \text{ovr} = C_{n-1} \oplus C_n.$

On the other hand, $C_n = x_{n-1} \cdot y_{n-1} + C_{n-1} \cdot (x_{n-1} + y_{n-1})$, a relation demonstrated in problem 3.E.1. Therefore the relationship becomes:

$$\text{ovr} = C_{n-1} \oplus [C_{n-1} \cdot (x_{n-1} + y_{n-1}) + x_{n-1} \cdot y_{n-1}].$$

This is of the form $\quad \alpha \oplus [\alpha \cdot \beta + \gamma]$, hence:

$$\bar{\alpha} \cdot (\alpha\beta + \gamma) + \alpha \cdot (\bar{\alpha} + \bar{\beta}) \cdot \bar{\gamma} = \bar{\alpha} \cdot \beta + \alpha \cdot \bar{\beta} \cdot \bar{\gamma}$$

From this: $\text{ovr} = \overline{C_{n-1}} \cdot x_{n-1} \cdot y_{n-1} + C_{n-1} \cdot \bar{x}_{n-1} \cdot \overline{y_{n-1}} \cdot (\overline{x_{n-1}} + \overline{y_{n-1}})$

$$= x_{n-1} \cdot y_{n-1} \cdot \overline{C_{n-1}} + \overline{x_{n-1}} \cdot \overline{y_{n-1}} \cdot C_{n-1}.$$

Now: $r_{n-1} = x_{n-1} \oplus y_{n-1} \oplus C_{n-1} \Rightarrow C_{n-1} = x_{n-1} \oplus y_{n-1} \oplus r_{n-1}.$

Substituting:

$$\begin{aligned}
\text{ovr} &= x_{n-1} \cdot y_{n-1} \cdot \overline{x_{n-1} \oplus y_{n-1} \oplus r_{n-1}} \\
&\quad + \overline{x_{n-1}} \cdot \overline{y_{n-1}} \cdot (x_{n-1} \oplus y_{n-1} \oplus r_{n-1}) \\
&= (x_{n-1} \oplus y_{n-1} \oplus \overline{r_{n-1}}) \cdot x_{n-1} \cdot y_{n-1} \\
&\quad + (x_{n-1} \oplus y_{n-1} \oplus r_{n-1}) \cdot \overline{x_{n-1}} \cdot \overline{y_{n-1}} \\
&= x_{n-1} \cdot y_{n-1} \oplus x_{n-1} \cdot y_{n-1} \oplus x_{n-1} \cdot y_{n-1} \cdot \overline{r_{n-1}} \\
&\quad + \bar{x}_{n-1} \cdot \bar{y}_{n-1} \cdot r_{n-1},
\end{aligned}$$

from which, noticing that $x_{n-1} \cdot y_{n-1} \oplus x_{n-1} \cdot y_{n-1} = 0$:

$$\text{ovr} = x_{n-1} \cdot y_{n-1} \cdot \overline{r_{n-1}} + \overline{x_{n-1}} \cdot \overline{y_{n-1}} \cdot r_{n-1}.$$

This relation does not depend on C_{n-1}, which does not appear in the output of the set of adders. It shows clearly the two following cases:

— overflow occurs if $NX^* > 0$ and $NY^* > 0$, that is $x_{n-1} = 0$ and $y_{n-1} = 0$; the result is negative, $r_{n-1} = 1$, and ovr $= 1$;
— overflow also occurs if $NX^* < 0$ and $NY^* < 0$, that is $x_{n-1} = 1$ and $y_{n-1} = 1$; the result is positive, $r_{n-1} = 0$, and ovr $= 1$.

Consequently, if ovr $= 0$, the relation $NR^* = NX^* + NY^* + C_0$ shows that a binary adder is suitable.

5.S.6.2 Subtraction

1. The operation $NR^* = NX^* - NY^* - C_0$, can be written:

$$NR^* = NX^* + (-NY^*) - C_0.$$

Now: $NY^* = y_{n-1}2^n + NY$, from which:

$$NR^* = NX^* + (y_{n-1}2^n - NY) - C_0.$$

Two cases can be envisaged depending on the value of y_{n-1}.
— If $y_{n-1} = 1$, that is to say $NY^* < 0$:

$$NR^* = NX^* + (2^n - NY) - C_0.$$

Let us put $NY_C = 2^n - NY$. This is called the 2's complement of NY. It gives:

$$NR^* = NX^* + NY_C.$$

Now $NY_C = 2^n - NY = (2^n - 1 - NY) + 1$, and we have seen on page 151 that $2^n - 1 - NY = N\bar{Y}$, from which:

$$\begin{aligned} NR^* &= NX^* + N\bar{Y} + 1 - C_0 \\ &= NX^* + N\bar{Y} + \bar{C}_0. \end{aligned}$$

Example: $n = 5$

let $X = 01001$, from which $NX = 9$ and $NX^* = +9$
$Y = 11011$, from which $NY = 27$ and $NY^* = -5$

Hence, $NY_C = 2^5 - 27 = 5 \Rightarrow Y_C = 00101$.
That is, $\bar{Y} = 00100$ and $Y_C = 00100 + 1 = 00101$.
The adder will realize:

$$\begin{array}{r} 1 \\ 01001 \\ +00100 \\ \hline 01110 \end{array} \Rightarrow +14 \qquad\qquad \begin{array}{r} 0 \\ 01001 \\ +00100 \\ \hline 01101 \end{array} \Rightarrow +13$$

if $C_0 = 0$ if $C_0 = 1$

that is

$+9 - (-5) - 0 = +14$ $+9 - (-5) - 1 = +13$

— if $y_{n-1} = 0$, that is to say $NY^* > 0$:

$$NR^* = NX^* - NY - C_0 = NX^* + (-NY) - C_0$$

Let us put $-NY = NY_C^* = -2^n + NY_C \Rightarrow NY_C = 2^n - NY$. NY_C is the 2's complement of NY, and we can write $NY_C = N\bar{Y} + 1$. It gives:

$$NR^* = NX^* - 2^n + N\bar{Y} + 1 - C_0$$
$$= NX^* + N\bar{Y} + \bar{C}_0 - 2^n.$$

We have returned to the preceding case with the condition that the carry beyond the sign must be ignored.

Example: $n = 5$

$X = 01001$, from which $NX = 9$ and $NX^* = +9$

$Y = 00101$, from which $NY = 5$ and $NY^* = +5$.

Hence, $NY_C = 2^5 - 5 = 27 \Rightarrow Y_C = 11011$.

That is $\bar{Y} = 11010$ and $Y_C = 11010 + 1 = 11011$.
The adder will realize:

1	0
01001	01001
+ 11010	+ 11010
1 00100 ⇒ +4	1 00011 ⇒ +3
↑ carry ignored	↑ carry ignored

that is

$+9 - (+5) - 0 = +4$ $\qquad$ $+9 - (+5) - 1 = +3$

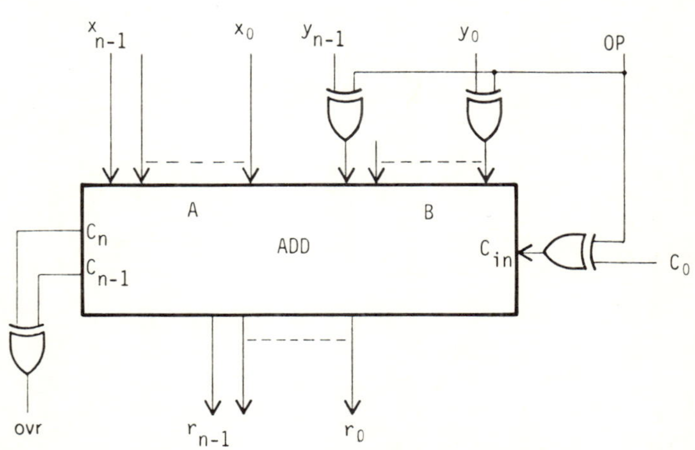

Figure 5.55

2. The diagram may easily be established. It is given in Figure 5.55.

We notice that Y and C_0 are transmitted directly if $\mathrm{OP} = 0$, for the case of addition, but complemented if $\mathrm{OP} = 1$, for the case of subtraction.

5.S.7. Ternary matrix arithmetic

1. The tables of Figure 5.56 are easily obtained from the corresponding operations (Figure 5.57).

$b_n \backslash a_n$	2	1	0
0	2	1	0
1	0	2	1
2	1	0	2

$b_n \backslash a_n$	2	1	0
0	0	0	0
1	1	0	0
2	1	1	0

Figure 5.56

$$\begin{array}{c|ccc} & 0 & 1 & 0 \\ + & 0 & 0 & 1 \\ \hline 0 & 0 & 1 & 0 \\ \uparrow & & \uparrow & \\ \text{carry} & & \text{sum} & \end{array} \qquad \begin{array}{c|ccc} & 1 & 0 & 0 \\ + & 0 & 1 & 0 \\ \hline 1 & 0 & 0 & 1 \end{array} \qquad \begin{array}{c|ccc} & 1 & 0 & 0 \\ + & 1 & 0 & 0 \\ \hline 1 & 0 & 1 & 0 \end{array}$$

Figure 5.57

Notice that we have operated in the order:

$$\begin{array}{r} 1 \\ +0 \\ \hline 1 \end{array} \qquad \begin{array}{r} 2 \\ +1 \\ \hline 3 \end{array} \qquad \begin{array}{r} 2 \\ +2 \\ \hline 4 \end{array}$$

2a. Notice that the indices of the binary variables recall the weight of each variable:

	a_2	a_1	a_0
0 is coded by	0	0	1
1 is coded by	0	1	0
2 is coded by	1	0	0

Let us consider the tables of Figure 5.56 and draw the two corresponding matrices: Figure 5.58 is obtained.

At the intersections, we have recalled the values from the tables. Thus the diagonal corresponds to the value 2; bit S'_2 therefore will be in the 1 state when a_2 and b_0 or a_1 and b_1, or a_0 and b_2 are at 1.

Similar reasoning defines S'_1, S'_0 and C'_{n+1}.

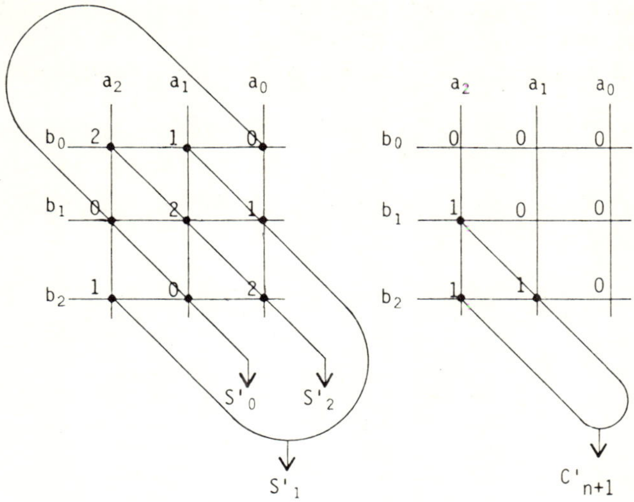

Figure 5.58

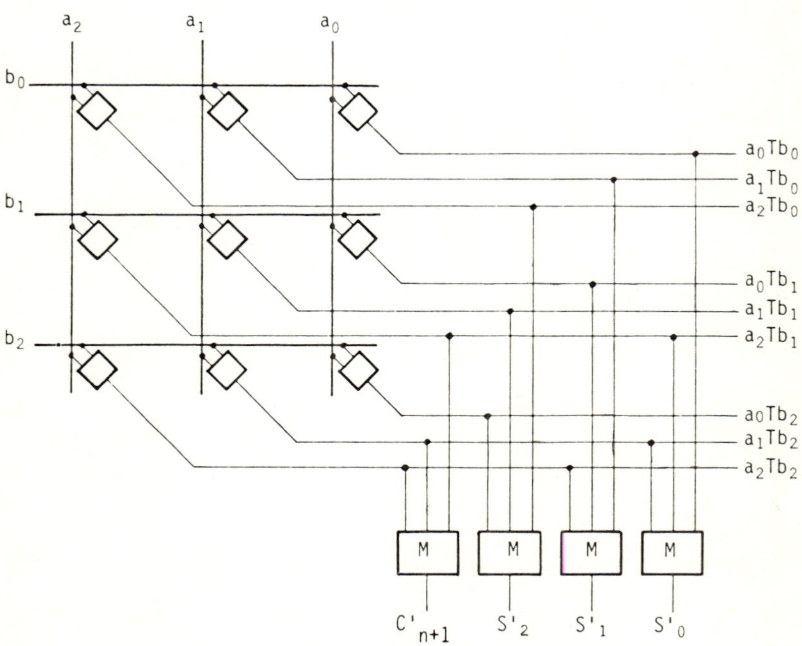

Figure 5.59

2b. It follows that the T operations are AND and the M operations are OR (Figure 5.59).

2c. The following equations can be deduced:

$$S'_0 = a_0 \cdot b_0 + a_2 \cdot b_1 + a_1 \cdot b_2 \qquad S'_1 = a_1 \cdot b_0 + a_0 \cdot b_1 + a_2 \cdot b_2,$$
$$S'_2 = a_2 \cdot b_0 + a_0 \cdot b_2 + a_1 \cdot b_1 \qquad C'_{n+1} = a_2 \cdot b_1 + a_1 \cdot b_2 + a_2 \cdot b_2.$$

3. The modified tables are given in Figure 5.60, together with the corresponding operations (Figure 5.61).

S_n:

b_n \ a_n	2	1	0
0	0	2	1
1	1	0	2
2	2	1	0

C_{n+1}:

b_n \ a_n	2	1	0
0	1	0	0
1	1	1	0
2	1	1	1

Figure 5.60

```
    0 1 0←Cn          0 1 0←Cn          0 1 0←Cn
    --------          --------          --------
    0 1 0             1 0 0             1 0 0
 +  0 0 1          +  0 1 0          +  1 0 0
 ---------         ---------         ---------
 0 | 1 0 0         1 | 0 1 0         1 | 1 0 0

    0 0 1←Cn          0 1 0←Cn          0 0 1←Cn
    --------          --------          --------
    0 1 0             0 0 1             0 0 1
 +  0 1 0          +  1 0 0          +  0 0 1
 ---------         ---------         ---------
 0 | 1 0 0         1 | 0 0 1         0 | 0 0 1
```

Figure 5.61

This permits the treatment of multi-digit ternary coded binary numbers:

```
   1 0 1 0←carries     → 0 1 0 | 0 0 1 | 0 1 0 | 0 0 1
   ---------             ------------------------------
   2 1 0 2               1 0 0 | 0 1 0 | 0 0 1 | 1 0 0
 + 1 2 1 1             + 0 1 0 | 1 0 0 | 0 1 0 | 0 1 0
 -----------           --------------------------------
 1 | 1 0 2 0           1 | 0 1 0 | 0 0 1 | 1 0 0 | 0 0 1
```

4. A comparable argument to that given in the answer to problem 2 enables the corresponding diagram for system carry $C_n = 1$ to be drawn. Consequently,

the required arrangement is the multiplex of the two diagrams considered previously for each value of C_n (Figure 5.62), with a simplification for C_{n+1}.

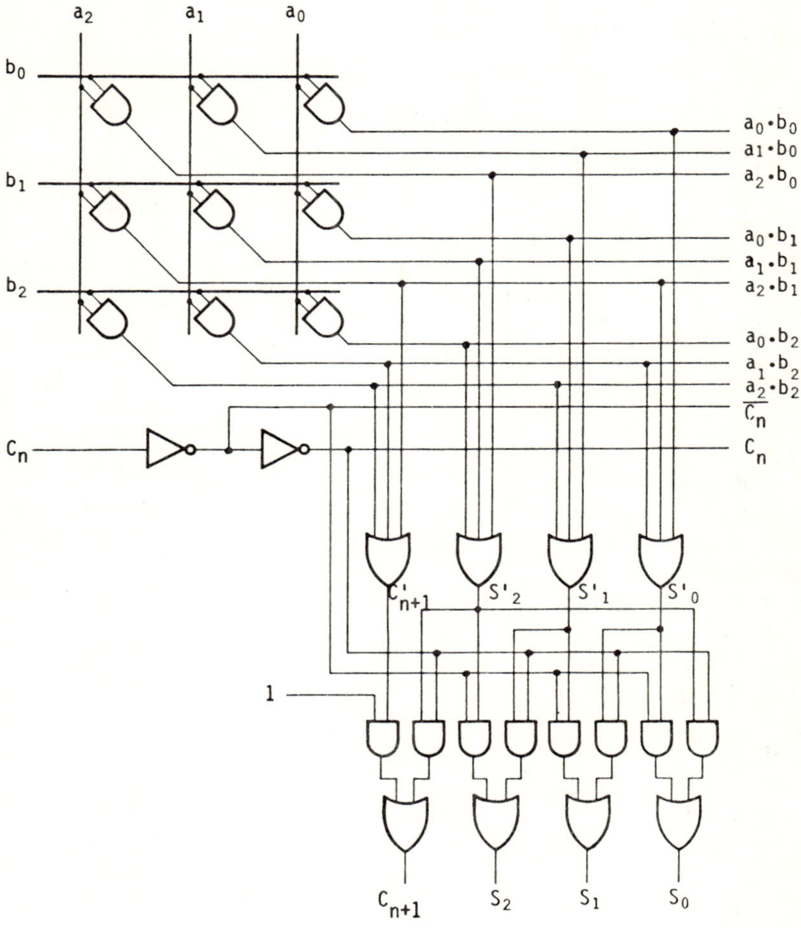

Figure 5.62

5. The following relations can easily be deduced:

$$S_0 = \overline{C_n}\cdot(a_0\cdot b_0 + a_2\cdot b_1 + a_1\cdot b_2) + C_n\cdot(a_2\cdot b_0 + a_0\cdot b_2 + a_1\cdot b_1)$$

$$S_1 = \overline{C_n}\cdot(a_1\cdot b_0 + a_0\cdot b_1 + a_2\cdot b_2) + C_n\cdot(a_0\cdot b_0 + a_2\cdot b_1 + a_1\cdot b_2)$$

$$S_2 = \overline{C_n}\cdot(a_2\cdot b_0 + a_0\cdot b_2 + a_1\cdot b_1) + C_n\cdot(a_1\cdot b_0 + a_0\cdot b_1 + a_2\cdot b_2)$$

$$C_{n+1} = (a_2\cdot b_1 + a_1\cdot b_2 + a_2\cdot b_2) + C_n\cdot(a_2\cdot b_0 + a_0\cdot b_2 + a_1\cdot b_1)$$

For $C_n = 0$, we take the results given in answer 2c. For $C_n = 1$, there is an offset indicated by the arrows.

6a. The diagram for a connection of four modules is given in Figure 5.63.

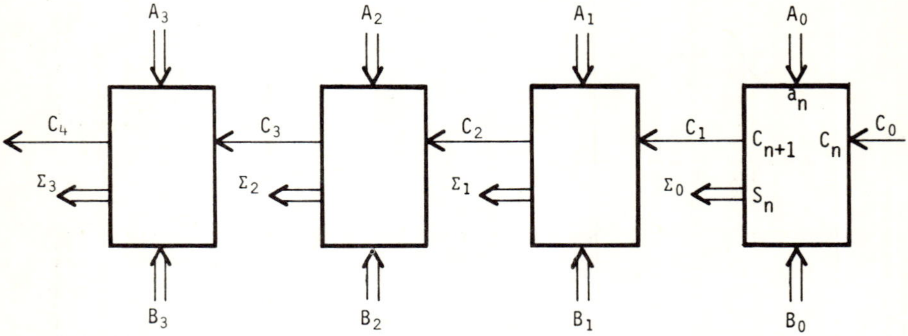

Figure 5.63

6b. The diagram of Figure 5.62 shows that C_{n+1}, S_2, S_1, and S_0 are obtained from four levels of logic, hence a calculation time of 40 ns. Since the modules are in series from the point of view of carry propagation, C_4 is obtained after $40 \times 4 = 160$ ns.

6c. The expression for C_{n+1} shows that if we put:

$$\delta_n = a_2 \cdot b_1 + a_1 \cdot b_2 + a_2 \cdot b_2 \qquad \text{and} \qquad \gamma_n = a_2 \cdot b_0 + a_0 \cdot b_2 + a_1 \cdot b_1,$$

the carry can be writtein:

$$C_{n+1} = \delta_n + \gamma_n \cdot C_n.$$

6d. The relations are the following:

$$\begin{aligned} C_1 &= \delta_0 + \gamma_0 \cdot C_0 \\ C_2 &= \delta_1 + \gamma_1 \cdot C_1 = \delta_1 + \gamma_1 \cdot \delta_0 + \gamma_1 \cdot \gamma_0 \cdot C_0 \\ C_3 &= \delta_2 + \gamma_2 \cdot C_2 = \delta_2 + \gamma_2 \cdot \delta_1 + \gamma_2 \cdot \gamma_1 \delta_0 + \gamma_2 \cdot \gamma_1 \cdot \gamma_0 \cdot C_0 \\ C_4 &= \delta_3 + \gamma_3 \cdot C_3 = \delta_3 + \gamma_3 \cdot \delta_2 + \gamma_3 \cdot \gamma_2 \cdot \delta_1 \\ &\qquad + \gamma_3 \cdot \gamma_2 \cdot \gamma_1 \cdot \delta_0 \\ &\qquad\qquad + \gamma_3 \cdot \gamma_2 \cdot \gamma_1 \cdot \gamma_0 \cdot C_0. \end{aligned}$$

6e. Problem *5.E.3* allows us to write:

$$\overline{G_0} = \overline{\delta_0},\ \overline{P_0} = \overline{\gamma_0}; \overline{G_1} = \overline{\delta_1}, \overline{P_1} = \overline{\gamma_1};$$
$$\overline{G_2} = \overline{\delta_2}, \overline{P_2} = \overline{\gamma_2}; \overline{G_3} = \overline{\delta_3}, \overline{P_3} = \overline{\gamma_3};$$
$$C_{n+x} = C_1, \quad C_{n+y} = C_2, \quad C_{n+z} = C_3.$$

6f. It is sufficient to provide the matrix adders with outputs $\overline{\delta_n}$ and $\overline{\gamma_n}$; thus the diagram of Figure 5.64 is obtained.

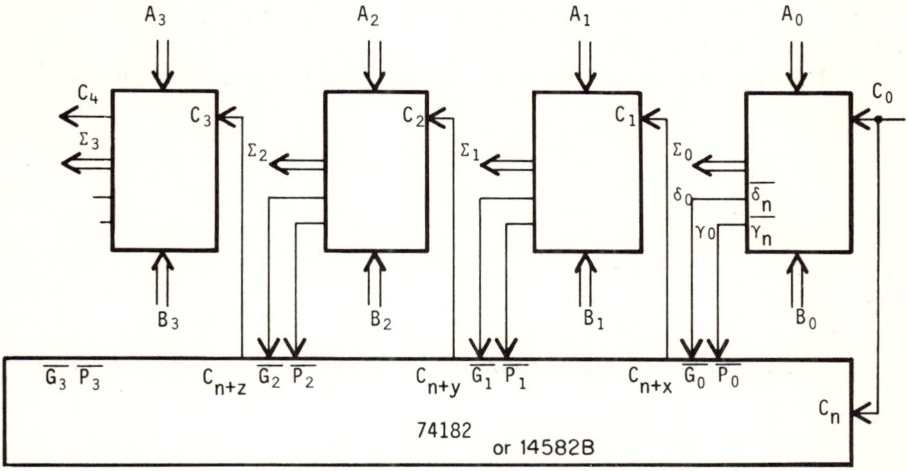

Figure 5.64

The calculation time is defined by Figure 5.65, which uses the elements of Figure 5.42. We obtain a value half of that obtained for the structure of Figure 5.63.

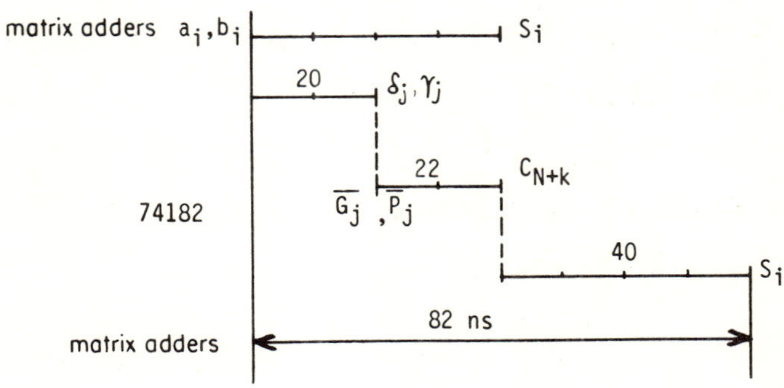

Figure 5.65

6. *Digital systems*

6.1 SUMMARY OF THEORY

The specification of an application for a digital system can be represented by a PETRINET, in the case of industrial automation, or by a *flow chart*.

The flow chart, which describes the algorithm used to solve a problem, includes actions to be carried out, which we shall call *operations*, and *tests*, logical choices between two exits which allows internal or extrenal events to intervene in the operation of the system. Operations and tests are represented by *rectangles* and *diamond shapes* respectively.

The digital system under consideration must therefore have an architecture which allows:

(a) management of the flow chart, that is the routing of all its branches;
(b) execution of actions encountered.

This architecture can, therefore, be partitioned into:

(a) a *monitoring and control unit* which ensures the management of the flow chart;
(b) a *processing unit* which ensures the executor of the actions.

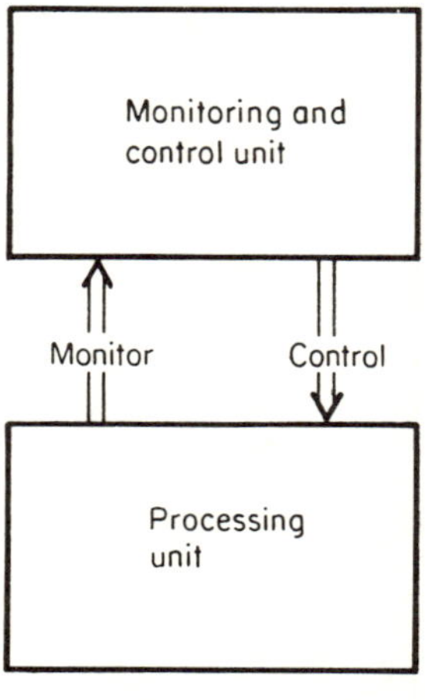

Figure 6.1

The arrangement is represented in Figure 6.1. This block diagram recalls that of industrial process control by a data processing machine: computer, mini-, or microcomputer, microprogrammable logic or wired logic. Such a machine is represented in Figure 6.2.

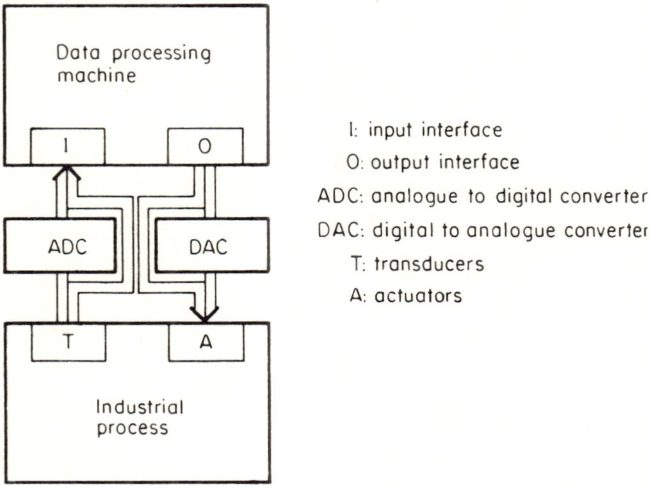

Figure 6.2

6.2 STATEMENT OF EXERCISES AND PROBLEMS

Synthesis of the processing unit

Its architecture depends essentially on the nature of the operations which appear on the flow chart. However, a certain number of *elementary functions* can be established from which a judicious arrangement allows the required processing unit to be realized.

6.E.1 Elementary functions

Arithmetic and logic unit (ALU)

This allows the following operations between two words: binary addition and subtraction, OR, AND, exclusive OR, equality detection.

It is assembled from 74181 or 14581B integrated circuits as studied in problem 5.E.2. A reperesentation is given in Figure 6.3.

1. Give the schematic diagram of an ALU for the processing of n bits according to the table of Figure 6.4.

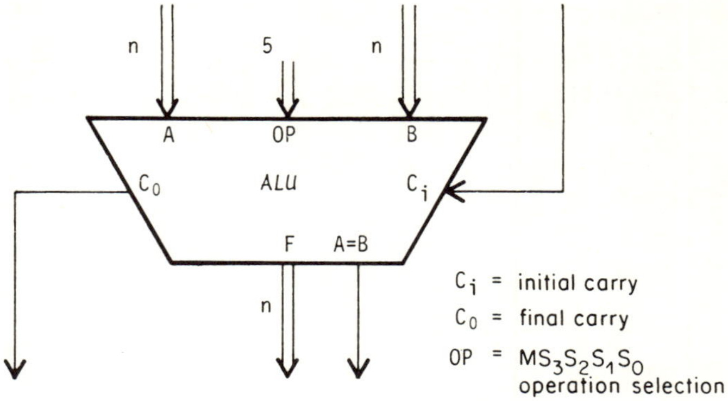

Figure 6.3

OP	operation
09	$2^n C_0$ plus $NF = NA$ plus NB plus C_i
06	$2^n C_0$ plus $NF = NA - NB - C_i$
1E	$F = A + B$
1B	$F = A \cdot B$
16	$F = A \oplus B$
19	$F = A \odot B$
1F	$F = A$
1A	$F = B$
13	$F = 0$
1C	$F = 1$

Figure 6.4

2-input multiplexer (MUX)

This allows the selection of one word from two. It is assembled from 74157 or 14053B integrated circuits. A repesentation is given in Figure 6.5.

2. Give the schematic diagram of a MUX for the processing of n bits according to the table of Figure 6.6.

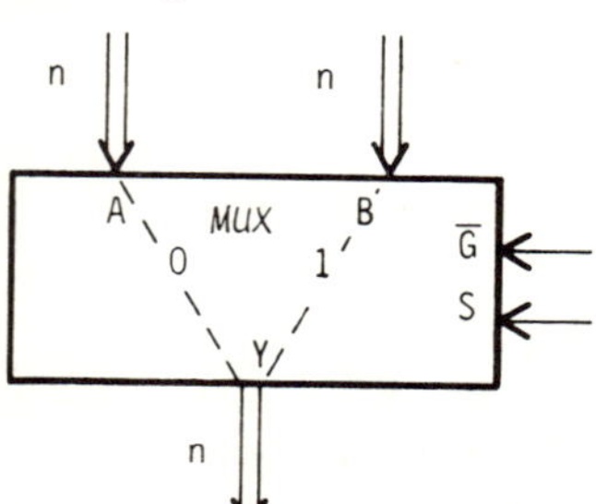

Figure 5.5

$\bar{G}$	S	operation
0	0	$Y = A$
0	1	$Y = B$
1	—	$Y = 0$

Figure 6.6

Conditional complementer (CC)

This allows a word to be complemented, or not, according to the state of a control signal. It is assembled from 7486 or 14070B integrated circuits. A representation is given in Figure 6.7.

3. Give the schematic diagram of a CC for the processing of n bits according to the table of Figure 6.8.

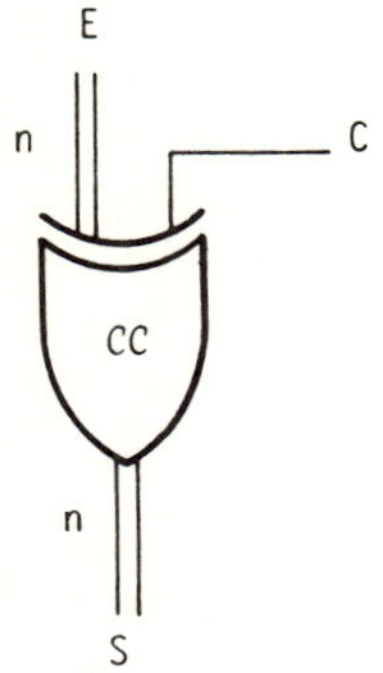

Figure 6.7

C	operation
0	$S = E$
1	$S = \bar{E}$

Figure 6.8

Gate (G)

This allows the transmission, or not, of a word according to the state of a control signal. It is assembled from 7408 or 14081B integrated circuits. A representation is given in Figure 6.9.

4. Give the schematic diagram of a gate G for the processing of n bits according to the table of Figure 6.10.

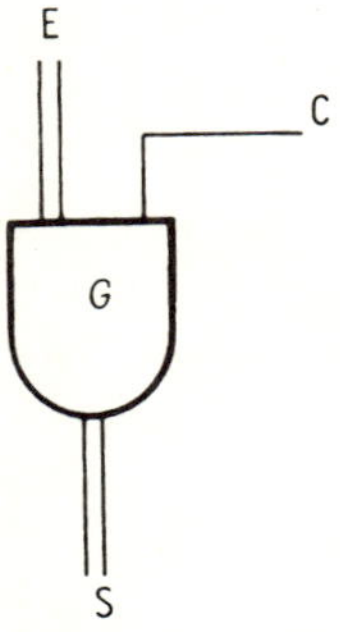

Figure 6.9

C	*operation*
0	$S = 0$
1	$S = E$

Figure 6.10

Flip-flop (FF)

This allows the storage of one bit when the clock input is activated. A distinction is made between the D flip-flop, represented in Figure 6.11, and the JK flip-flop, represented in Figure 6.12.

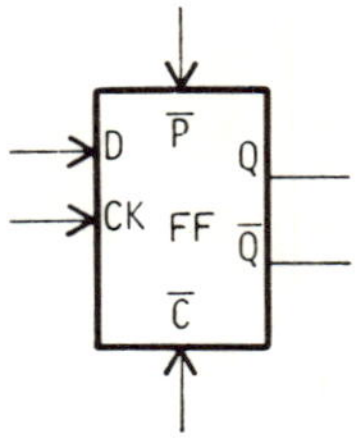

$\bar{P}$	$\bar{C}$	D	CK	Q
0	1	—	—	1
1	0	—	—	0
1	1	0	↑	0
1	1	1	↑	1

Figure 6.11

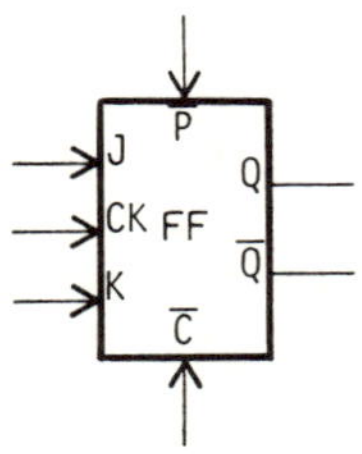

$\bar{P}$	$\bar{C}$	J	K	CK	Q
0	1	—	—	—	1
1	0	—	—	—	0
1	1	0	0	⎍	Q_0
1	1	0	1	⎍	0
1	1	1	0	⎍	1
1	1	1	1	⎍	$\bar{Q}_0$

Q_0 is the state of Q before activation of the close CK

Figure 6.12

Storage register (RSTO)

This allows the storage of a word, generally on the active edge of a clock. It is constructed from 74175 or 14175B integrated circuits. A representation is given in Figure 6.13.

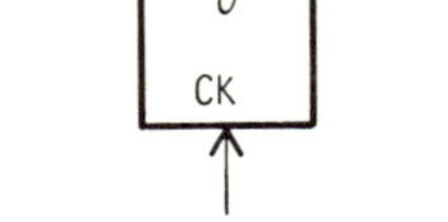

$\bar{C}$	D_i	CK	Q_i
0	—	—	0
1	0	↑	0
1	1	↑	1

Figure 6.13

Universal shift register (SR)

This allows storage and also shifting of a word to left or right when the clock input is activated. It is constructed from 74194 or 14194B integrated circuits. A representation is given in Figure 6.14.

5. Give the schematic diagram of a universal shift register for the processing of n bits according to the table of Figure 6.15. One constraint must be observed: S_1 and S_0 must not be changed when input CK is low.

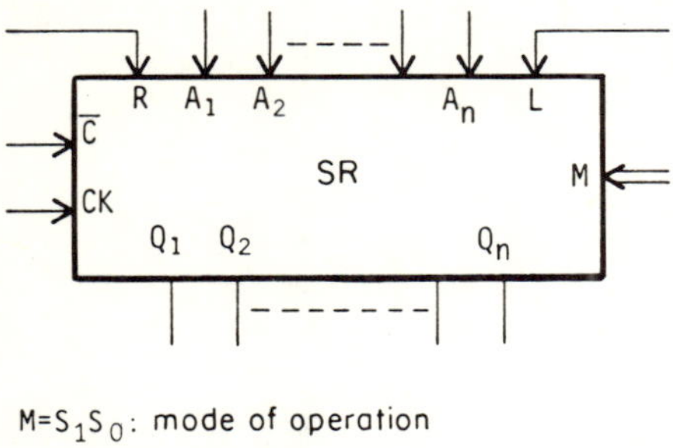

Figure 6.14

M	$\bar{C}$	CK	Q_1	Q_2	—
—	0	—			
0	1	↑	Q_{1_0}	Q_{2_0}	Q_{n_0}
1	1	↑	R	Q_1	Q_{n-1}
2	1	↑	Q_2	Q_3	L
3	1	↑	A_1	A_2	A_n

Figure 6.15

Reversible counter (CTR)

This allows a previously loaded word to be continuously incremented or decremented in synchronism with a clock signal. It is assembled from 74193 integrated circuits. A representation is given in Figure 6.16.

6. Give the schematic diagram of a reversible counter for the processing of n bits according to the table of Figure 6.17.

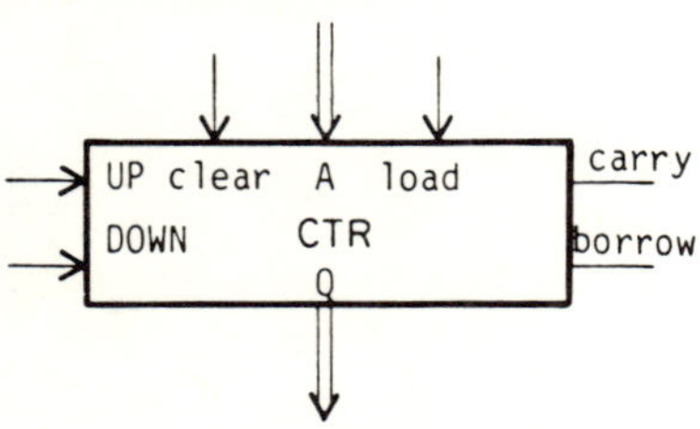

Figure 6.16

clear	load	UP	DOWN	Q
0	1	1	1	Q_0
1	—	—	—	0
0	0	—	—	A
0	1	↑	1	$Q_0 + 1$
0	1	1	↑	$Q_0 - 1$

Figure 6.17

Comparator (COMP)

This allows the comparison of two n bit words to detect whether they are equal or, if the two represent coded quantities, their relative magnitudes. It is assembled from 7485 or 14585B integrated circuits. A representation is given in Figure 6.18.

7. Give the schematic diagram of a comparator for the processing of n bits.

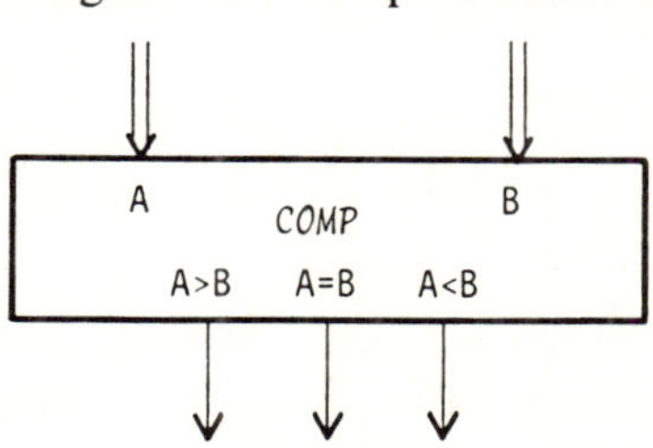

Figure 6.18

Incrementer decrementer (I/D)

This realizes the binary sum or difference of a word and a bit C according to the state of a control signal A. A representation is given in Figure 6.19.

8. Give a schemetic diagram of an ID for the processing of n bits according to the table of Figure 6.20.

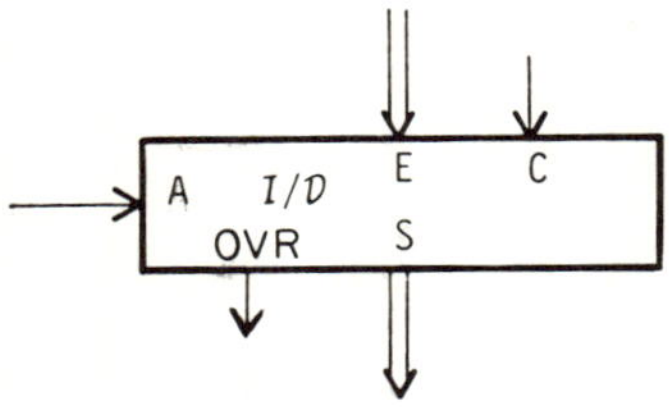

Figure 6.19

A	C	S
0	0	E
0	1	$E+1$
1	0	E
1	1	$E-1$

Figure 6.20

6.E.2 Binary divider

It is required to design a processing unit for synchronous 8-bit by 8-bit division of unsigned numbers. The operation is described by the flow chart of Figure 6.21, which corresponds to the method called *restoration of partial remainders.*

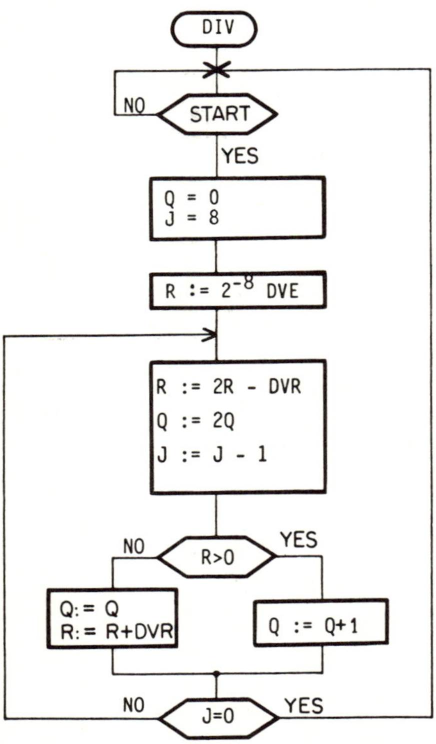

DVE is the dividend

DVR the divisor

Q the quotient

R the remainder

J indicates the end of the calculation

The symbol := signifies 'becomes'

Figure 6.21

Recall that $2^k A$ indicates that A is shifted k places to the left if $k > 0$, and k places to the right if k is negative.

For a clear explanation of the operation, consider an example (Figure 6.22),

C_8	R	Q	operation carried out
0	0 0	9 A	initialisation
	0 1	3 4	1 st shift
	- 2 B		SUB
1	D 6	3 (4)	LSBQ = 0
	+ 2 B		ADD
	0 1	3 4	
	0 2	6 8	2 nd shift
	- 2 B		SUB
1	D 7	6 (8)	LSBQ = 0
	+ 2 B		ADD
	0 2	6 8	
	0 4	D 0	3 rd shift
	- 2 B		SUB
1	D 9	D (0)	LSBQ = 0
	+ 2 B		ADD
	0 4	D 0	
	0 9	A 0	4 th shift
	- 2 B		SUB
1	D E	A (0)	LSBQ = 0
	+ 2 B		ADD
	0 9	A 0	
	1 3	4 0	5 th shift
	- 2 B		SUB
1	E 8	4 (0)	LSBQ = 0
	+ 2 B		ADD
	1 3	4 0	
	2 6	8 0	6 th shift
	- 2 B		SUB
1	F B	8 (0)	LSBQ = 0
	+ 2 B		ADD
	2 6	8 0	
	4 D	0 0	7 th shift
	- 2 B		SUB
0	2 2	0 (1)	LSBQ = 1
	4 4	0 2	8 th shift
	- 2 B		SUB
0	1 9	0 (3)	LSBQ = 1

Figure 6.22

0	0	1	2	3	4	5	6	7	8	9	A	B	C	D	E	F
0	0	1	2	3	4	5	6	7	8	9	A	B	C	D	E	F
1		2	3	4	5	6	7	8	9	A	B	C	D	E	F	10
2			4	5	6	7	8	9	A	B	C	D	E	F	10	11
3				6	7	8	9	A	B	C	D	E	F	10	11	12
4					8	9	A	B	C	D	E	F	10	11	12	13
5						A	B	C	D	E	F	10	11	12	13	14
6							C	D	E	F	10	11	12	13	14	15
7								E	F	10	11	12	13	14	15	16
8									10	11	12	13	14	15	16	17
9										12	13	14	15	16	17	18
A											14	15	16	17	18	19
B												16	17	18	19	1A
C													18	19	1A	1B
D														1A	1B	1C
E															1C	1D
F																1E

Figure 6.23

for which DVE = 9*A* and DVR = 2*B*. For ease of handling, this example has been developed with hexadecimal coded numbers.

R is a shift register, initialized to 00, which will contain the remainder at the end of the operation.

Q is a shift register, initialized to DVE, which will contain the quotient at the end of the operation.

C_8 is a flip-flop which accepts the carry generated by the subtraction.

In order to follow the operations carried out, the table of hexadecimal addition is recalled in Figure 6.23, the table for multiplication by 2 (for the shifts) in Figure 6.24, and the table for subtraction, $x - y$, in Figure 6.25.

0	1	2	3	4	5	6	7	8	9	A	B	C	D	E	F
0	2	4	6	8	A	C	E	10	12	14	16	18	1A	1C	1E

Figure 6.24

	0	1	2	3	4	5	6	7	8	9	A	B	C	D	E	F
0	0	1F	1E	1D	1C	1B	1A	19	18	17	16	15	14	13	12	11
1	1	0	1F	1E	1D	1C	1B	1A	19	18	17	16	15	14	13	12
2	2	1	0	1F	1E	1D	1C	1B	1A	19	18	17	16	15	14	13
3	3	2	1	0	1F	1E	1D	1C	1B	1A	19	18	17	16	15	14
4	4	3	2	1	0	1F	1E	1D	1C	1B	1A	19	18	17	16	15
5	5	4	3	2	1	0	1F	1E	1D	1C	1B	1A	19	18	17	16
6	6	5	4	3	2	1	0	1F	1E	1D	1C	1B	1A	19	18	17
7	7	6	5	4	3	2	1	0	1F	1E	1D	1C	1B	1A	19	18
8	8	7	6	5	4	3	2	1	0	1F	1E	1D	1C	1B	1A	19
9	9	8	7	6	5	4	3	2	1	0	1F	1E	1D	1C	1B	1A
A	A	9	8	7	6	5	4	3	2	1	0	1F	1E	1D	1C	1B
B	B	A	9	8	7	6	5	4	3	2	1	0	1F	1E	1D	1C
C	C	B	A	9	8	7	6	5	4	3	2	1	0	1F	1E	1D
D	D	C	B	A	9	8	7	6	5	4	3	2	1	0	1F	1E
E	E	D	C	B	A	9	8	7	6	5	4	3	2	1	0	1F
F	F	E	D	C	B	A	9	8	7	6	5	4	3	2	1	0

Figure 6.25

In Figure 6.22, the first to the sixth shifts inclusive provide a restoration of the partial remainder, since $C_8 = 1$ after subtraction; the least significant bit of the quotient, LSBQ, is set to 0.

The seventh and eighth shifts lead to non-restoration of the remainder since $C_8 = 0$; LSBQ is set to 1.

Finally Q = 03 and R = 19. That is:

$$(2B \times 03 + 19 = 9A)_{16}, \quad \text{or} \quad (43 \times 3 + 25 = 154)_{10}.$$

1. Give the block diagram of the processing unit.
2. Using the elementary functions reviewed in *problem 6.E.1*, give a detailed diagram and the associated model.

6.E.3 Transmission of characters

A memory contains a message to be transmitted to a machine called a teletype or TTY. The message is formed from a set of characters.

A character may be a number:0, 1,..., 9
a letter:A, B,..., Z
a sign :, ; ...
a symbol: $@
a command: line feed LF,
carriage return CR...

Each character is coded by 7 bits, so 128 are possible. They are generally represented by their equivalent hexadecimal value. The ASCII code is used as given in Figure 6.26.

Bits 4 to 6	—	0	1	2	3	4	5	6	7
	0	NUL	DLE	SP	0	@	P		p
	1	SOH	DC1	!	1	A	Q	a	q
	2	STX	DC2	″	2	B	R	b	r
	3	ETX	DC3	#	3	C	S	c	s
	4	EOT	DC4	$	4	D	T	d	t
	5	ENQ	NAK	%	5	E	U	e	u
	6	ACK	SYN	&	6	F	V	f	v
Bits 0 to 3	7	BEL	ETB	’	7	G	W	g	w
	8	BS	CAN	(	8	H	X	h	x
	9	HT	EM	)	9	I	Y	i	y
	A	LF	SUB	*	:	J	Z	j	z
	B	VT	ESC	+	;	K	[	k	{
	C	FF	FS	′	<	L	/	I	/
	D	CR	GS	-	=	M	]	m	}
	E	SO	RS	.	>	N	∧	n	≈
	F	SI	US	/	?	⊗	—	≠	DEL

Figure 6.26

The bits of each character are sent in synchronism with a clock, but the sequence of characters may be asynchronous. In order to separate the characters of the message, each character starts with a ‘0’, called the START bit, and finishes with two ‘1’s, called the STOP bits. A parity bit, even or odd, may follow the actual information.

The corresponding waveform is given in Figure 6.27.

Figure 6.28 represents, by way of example, the model of character K, coded by 4B with even parity.

The message to be transmitted is stored in a memory MEM. This has a parallel

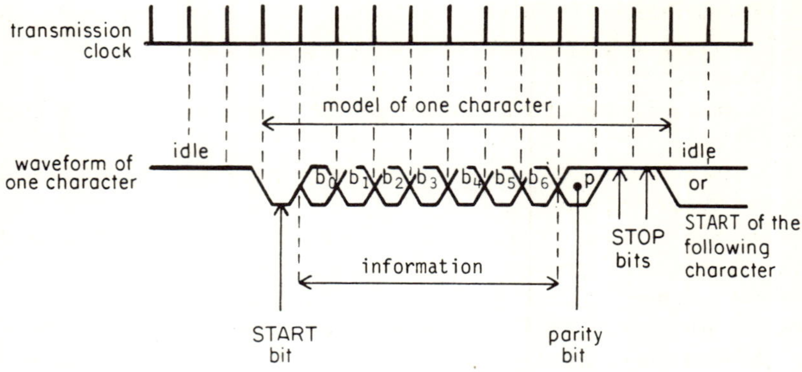

Figure 6.27

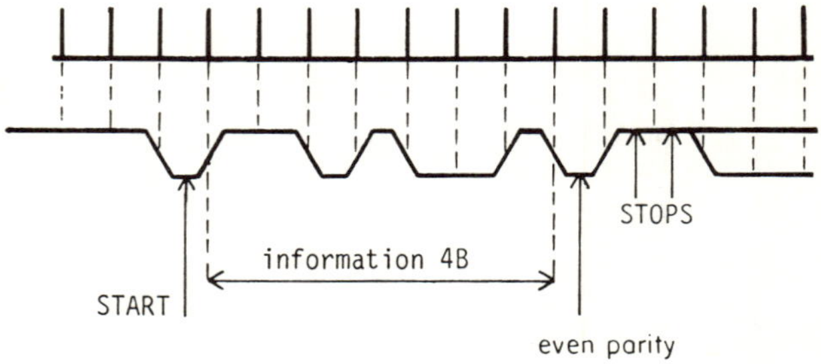

Figure 6.28

output; the bits of the model must be transmitted serially, so a shift register SR is interposed. Also, for the transmitted signal to be compatible with the teletype, an interface INT transforms a '0' to + 12 volts and a '1' to − 12 volts. Figure 6.29 gives a block diagram of the arrangement.

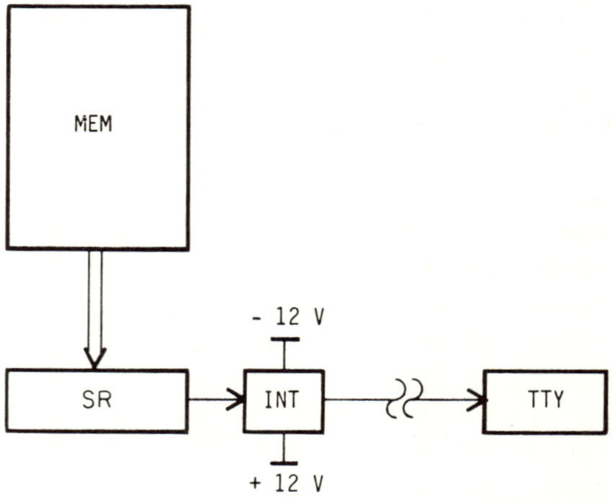

Figure 6.29

1a. Give a detailed block diagram, which uses the elementary functions defined in 6.E.1, taking account of the following points:

— The memory is loaded from another source;

— The end of the message is signalled by the character @ which will not be printed;

— The transmission clock has a frequency between 110 and 9 600 Hz. It is called the baud rate generator;

—A signal STA starts the transmission of the message. We suppose here that it arises from a briefly pressed button;

— An active high signal PWR indicates when power is applied;

— A generator provides the necessary clock phases.

1b. Give also the associated model.

2. Give the sequence diagram corresponding to the transmission of a message.

3. Give the memory contents corresponding to the following message:

```
␣␣FELIX␣QUI␣
            POTUIT␣RERUM
␣␣␣␣COGNOSCERE
           CAUSAS


␣␣␣VIRGILE␣␣␣(GEORGIQUES)
```

The sign ␣ indicates a space coded by 20.

4. Give the timing diagrams corresponding to the transmission of the message.

Synthesis of the monitoring and control unit

The architecture of this unit depends essentially on the manner in which it is decided to realize the flow chart. Having made the choice, it is necessary to develop the flow chart, which becomes a *state diagram.*

Several solutions are valid, based on the concepts of *wired logic* or *microprogrammed logic*. We have decided to design the control units for the two processes studied in *6.E.2* and *6.E.3* using the latter technique. In this case, the heart of the system is a ROM (read-only memory) or one of its derivatives: PROM, EPROM, etc.

Such a memory is characterized by an address word: $A_0, A_1, \ldots A_{p-1}$ which allows a data word $D_0, D_1, \ldots, D_{q-1}$ carrying the required information to be activated (see Figure 6.30).

Recall that such a memory is in fact a decoder: each output variable is a *canonical sum of products* of the address variables, and can therefore be simplified.

$$D_j = f(A_0, A_1, \ldots, A_{p-1}) \qquad \text{with } j \in 0, 1, \ldots, q-1$$

If this simplification is made, we obtain, *a posteriori*, the equations of the equivalent wired logic.

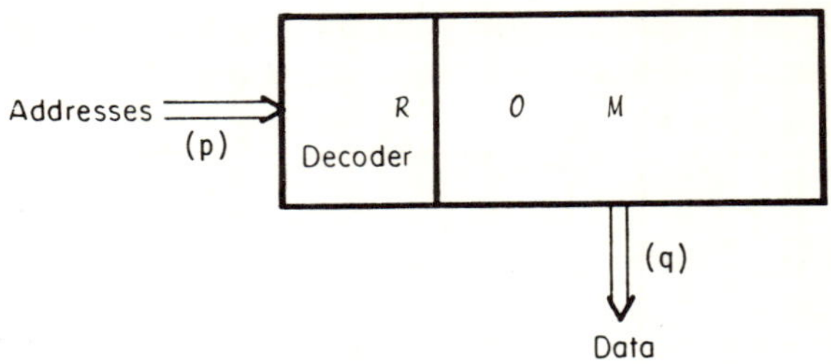

Figure 6.30

The state diagram must show the different steps of operation realized by *stages*, or *machine states* to which the *memory addresses* correspond. Among the possible solutions, we shall select two:

— The method called *state qualifier pair addressable*, realized by a control unit using micro-instructions and an address register, according to Figure 6.31.

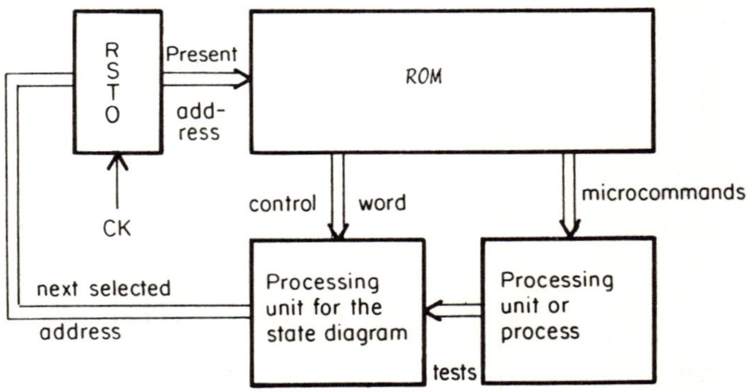

Figure 6.31

The micro-instruction has the format given in Figure 6.32.

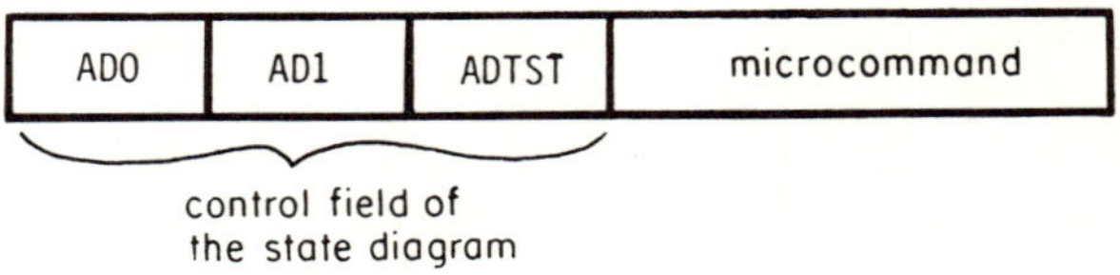

Figure 6.32

AD0 is the next address if the selected test is 0.

AD1 is the next address if the selected test is 1.

ADTST is the address of the selected test.

An address includes a group of compatible operations followed by a test. An application will be given in problem *6.E.5*.

— The method called *assumed address, variable format*, realized by a control unit using micro-instructions and an address counter, according to Figure 6.33.

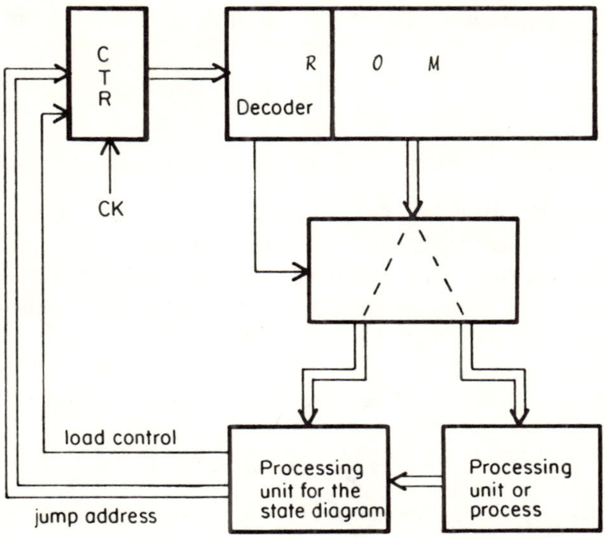

Figure 6.33

The micro-instruction has the double format given in Figure 6.34.

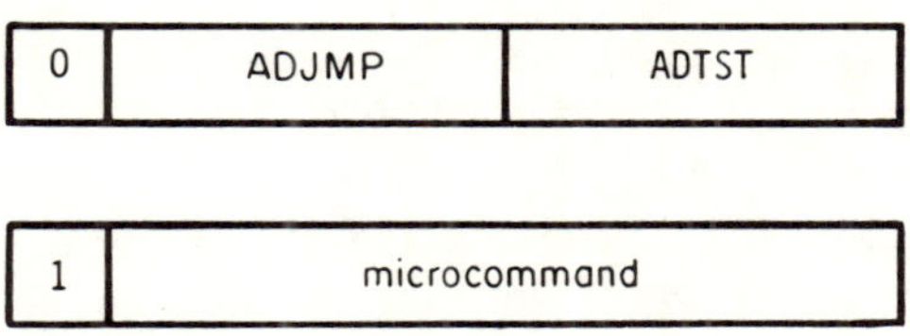

Figure 6.34

ADJMP is the jump address to be used.

ADTST is the address of the selected test.

The interpretation of the format is indicated by the first bit. Each group of compatible operations and each test has its own address. An application is given below.

6.E.4 Design of the control unit for the binary divider

This divider was the object of problem *6.E.2*

1. Give the flow diagram corresponding to the assumed address, variable format method. Take account of the following points:

— each test set to 1, leads to a test or following operation;

— each test set to 0 leads to a test or any operation;
— each operation leads to a test or the following operation.

This last point can include the creation of a fictitious test F always set to 0.

2. Using the elementary functions studied in problem *6.E.1*, give a detailed diagram. Specify the clock.

3. Give the microprogram to be loaded into memory.

6.E.5 Design of the control unit for the character transmission system

This system was the object of problem *6.E.3*.

1. The state diagram is given in Figure 6.49 (page 204). Verify that the constraints of the method are observed. Take account of the following point: when a state does not include a test, create a fictitious test F always set to 0.

2. Using the elementary functions studied in problem *6.E.1*, give a detailed diagram. Specify the clock.

3. Give the microprogram to be loaded into memory.

6.3 SOLUTIONS

6.S.1 Elementary functions

1. (ALU) The diagram is given in Figure 6.35. OP includes the inputs M, S_3, S_2, S_1, S_0 to which the weights 16, 8, 4, 2 and 1 have been assigned respectively. The $A = B$ outputs, of the open collector type, are connected together.

2. (MUX) The diagram is given in Figure 6.36.

3. (CC) The diagram is given in Figure 6.37.

4. (G) The diagram is given in Figure 6.38.

5. (SR) The diagram is given in Figure 6.39. The interconnections permit shifting to the right or left. M is formed by the inputs S_1 and S_0 to which the weights 2 and 1 have been assigned respectively.

6. (CTR) The diagram is given in Figure 6.40. It uses serial carry propagation.

7. (COMP) The diagram is given in Figure 6.41. It uses serial transmission of output data from the low to high weight digits.

8. (I/D) The diagram is given in Figure 6.42. It arises from the theory of adder-subtractors (see Chapter 5).

ADD	SUB
$S_K = a_K \oplus b_K \oplus C_K$	$S_K = a_K \oplus b_K \oplus C_K$
$C_{K+1} = a_K \cdot b_K + C_K \cdot (a_K + b_K)$	$C_{K+1} = \bar{a}_K \cdot b_K + C_K \cdot (\bar{a}_K + b_K)$

In our case, putting $b_K = 0$ and $a_K = E_K$, this becomes:

$$S_K = E_K \oplus C_K \qquad S_K = E_K \oplus C_K$$
$$C_{K+1} = E_K \cdot C_K \qquad C_{K+1} = \bar{E}_K \cdot C_K$$

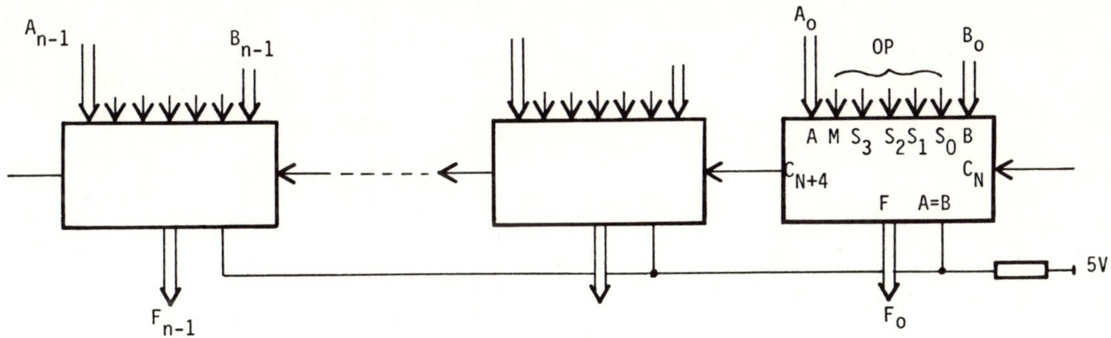

Figure 6.35

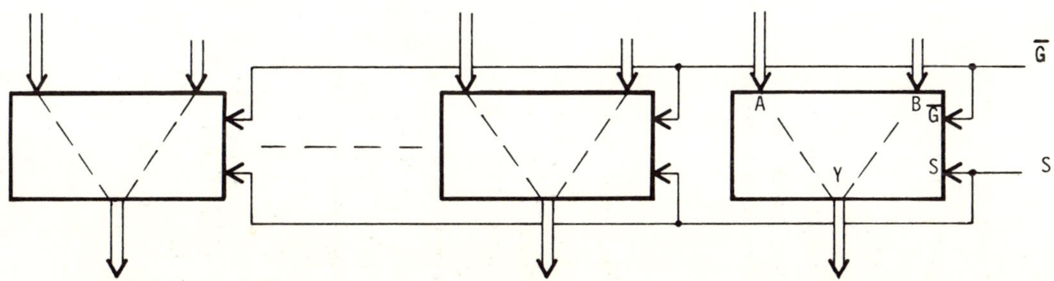

Figure 6.36

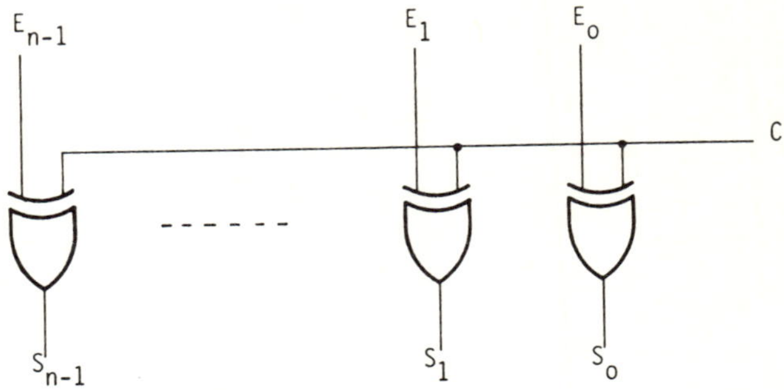

Figure 6.37

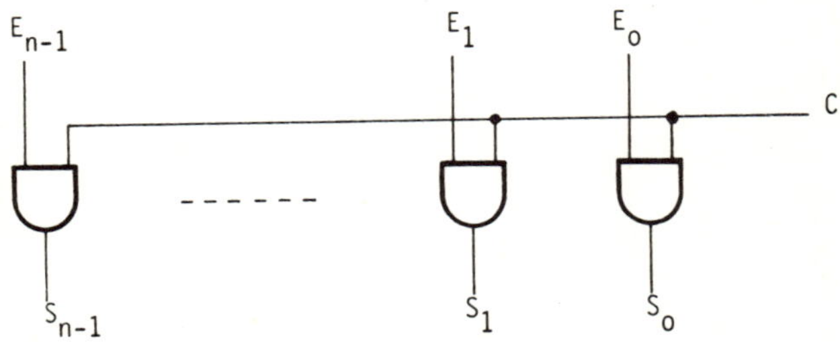

Figure 6.38

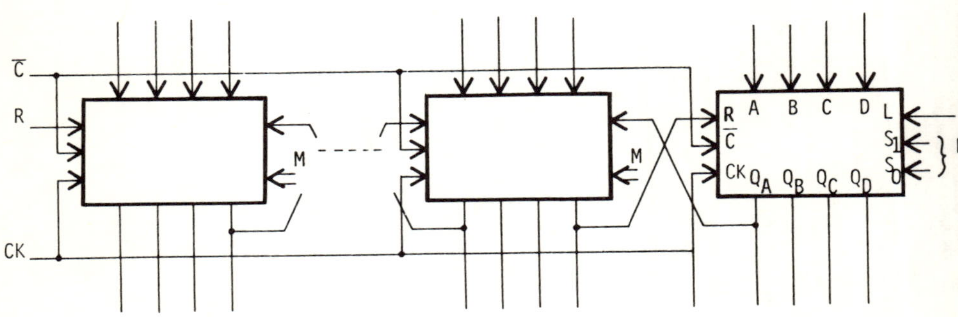

Figure 6.39

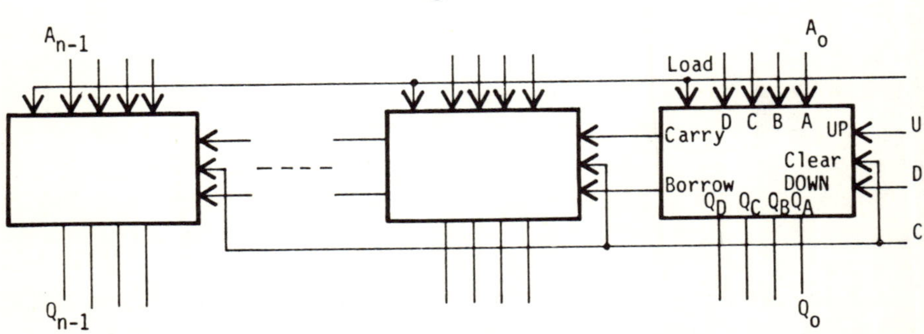

Figure 6.40

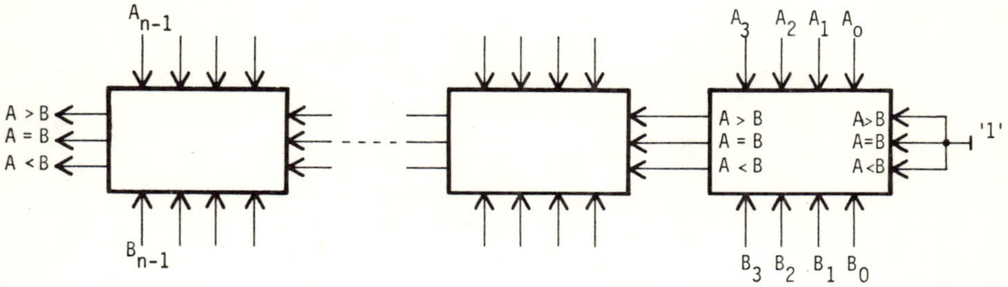

Figure 6.41

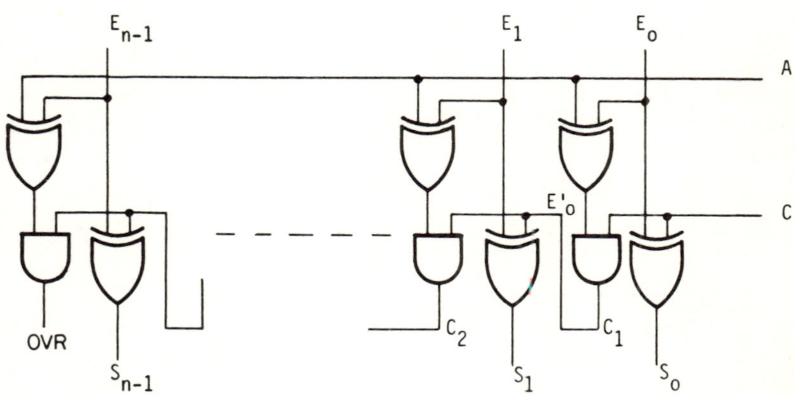

Figure 6.42

Putting $E'_K = E_K \oplus A$, we obtain: when

$$A = 0: E'_K = E_K \qquad \text{(ADD)}$$
$$A = 1: E'_K = \bar{E}_K \qquad \text{(SUB)}.$$

6.S.2 Binary divider

1. Examination of Figure 6.21 shows that R must contain the results of the successive sums or differences; the classical structure which allows this result to be obtained consists of feeding back the output of R to an ALU, the output of which feeds the input of R. In this case, R is called an *accumulator*. The corresponding diagram is given in Figure 6.43.

On the other hand, before effecting the sum or difference, it is necessary to shift (R) and (Q). The registers R and Q will therefore be shift registers (SR) and the register containing DVR will be a storage register (RSTO).

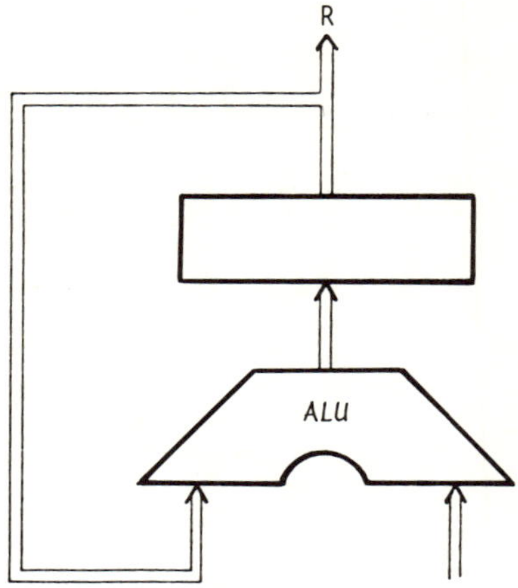

Figure 6.43

By adding a counter CTR in order to advance J, the end of operation indicator, a processing block to control the start of execution EX and a flip-flop FF containing the greatest weight of R, which allows the sign of R to be tested after each sum or difference: C_8; we obtain the diagram of Figure 6.44.

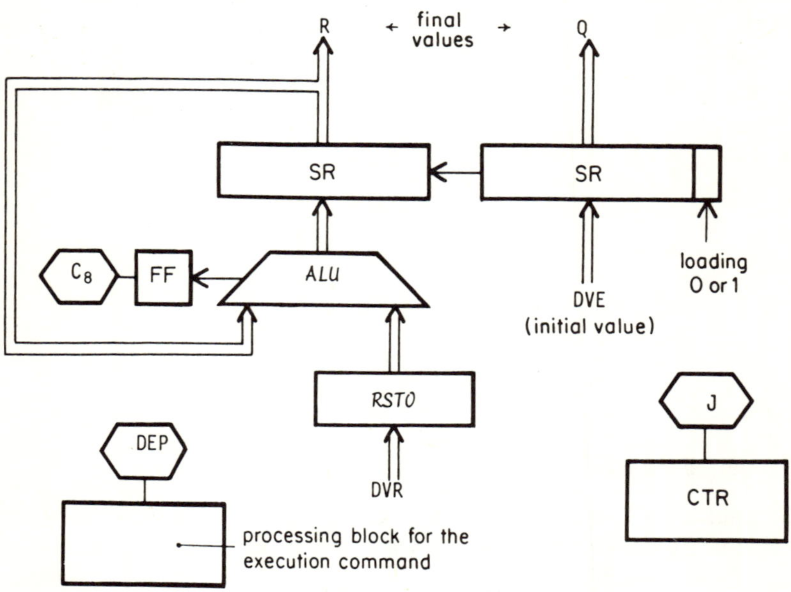

Figure 6.44

2. The detailed diagram is given in Figure 6.45 and the associated model in Figure 6.46. Bits $a, b, c, d, e, f, g, h, i, j, k$ will be determined by the control unit in order to obtain the actions described above; CK is a clock to be defined.

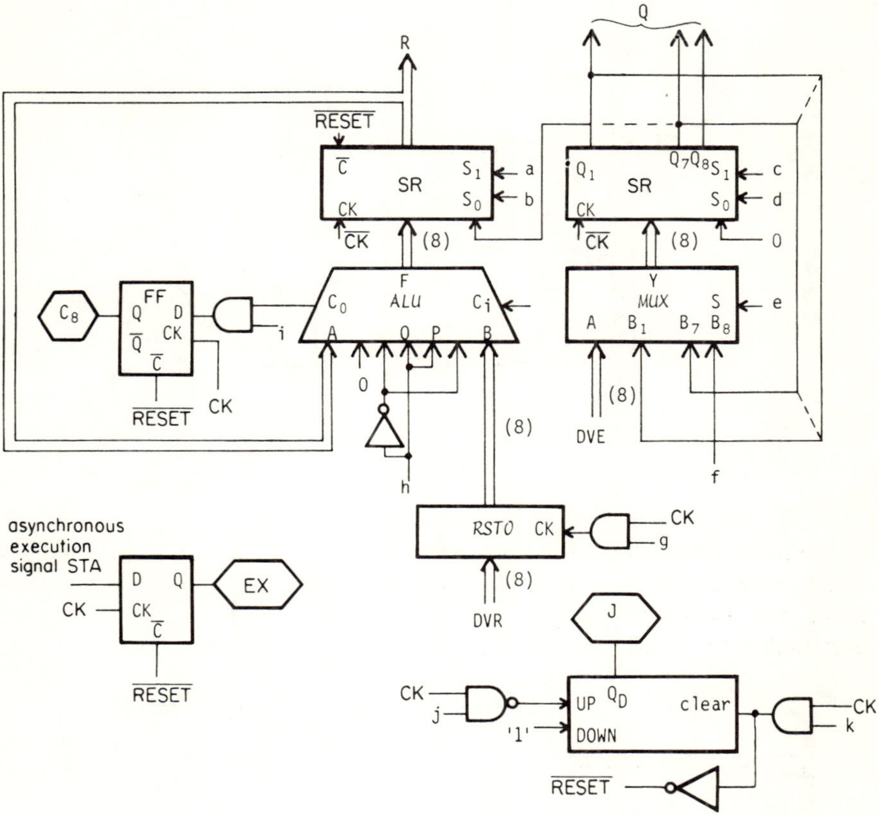

Figure 6.45

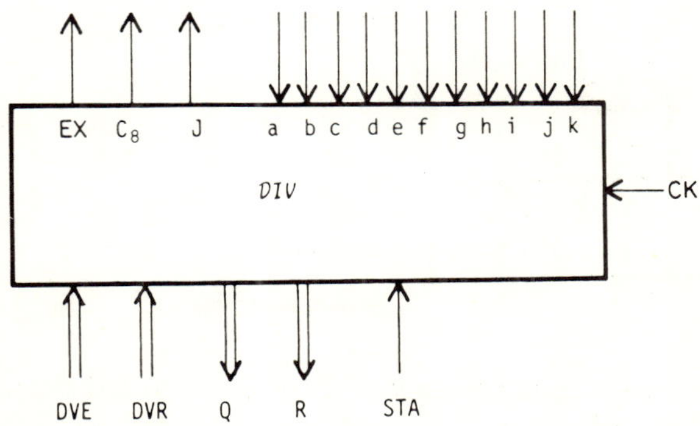

Figure 6.46

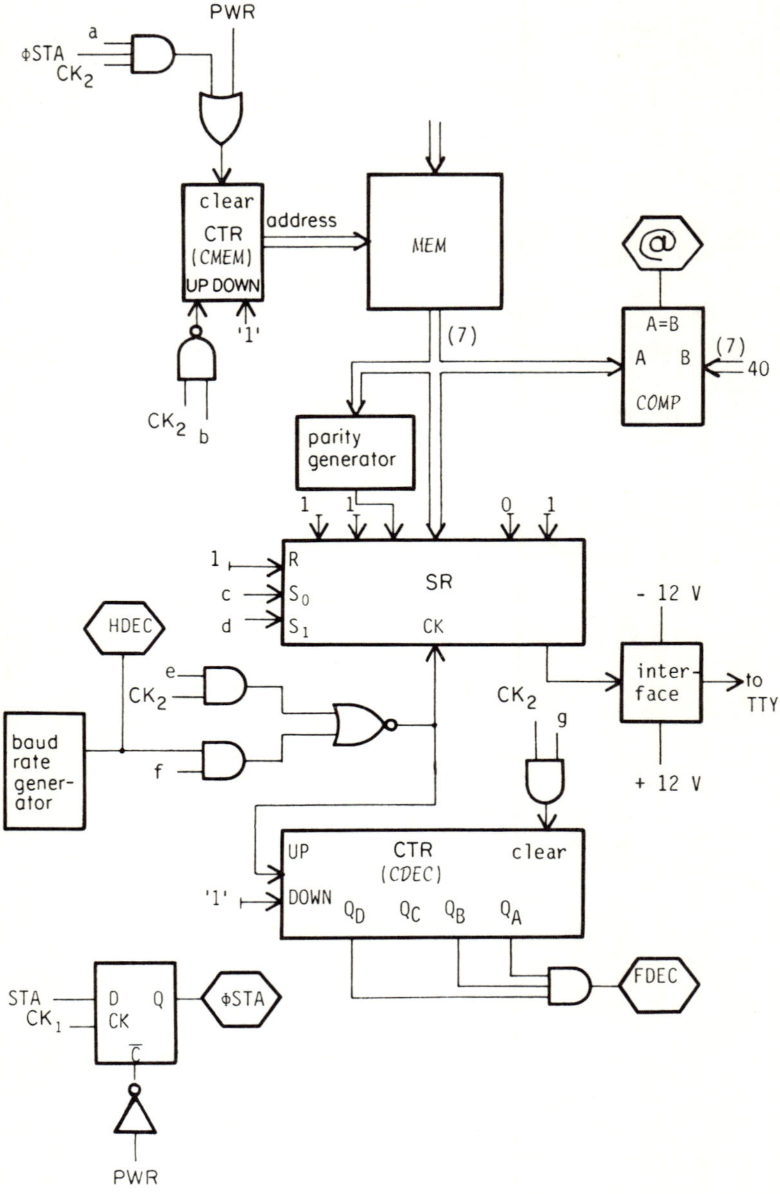
PWR
a
ΦSTA
CK_2
clear
CTR
(CMEM)
UP DOWN
address
MEM
'1'
CK_2
b
(7)
@
A=B
A
B
(7)
40
COMP
parity generator
1
1
0
1
1
R
c
S_0
d
S_1
SR
CK
- 12 V
HDEC
e
CK_2
inter-face
to TTY
CK_2
g
baud rate gener-ator
f
+ 12 V
UP
CTR
(CDEC)
clear
'1'
DOWN
Q_D
Q_C
Q_B
Q_A
FDEC
STA
CK_1
D
Q
CK
$\overline{C}$
ΦSTA
PWR

Figure 6.47

6.S.3 Transmission of characters

1a. The detailed diagram is given in Figure 6.47.

CK_1 and CK_2 are the two phases of the clock, as will be seen in the study of the control unit.

The message is stored in memory from address 0, provided by the counter CMEM.

When a character has been loaded into the shift register SR by a CK_2 pulse, the clock HDEC allows shifting of successive bits of the character to the output interface. At each shift, counter CDEC is incremented. When 11 bits have been transmitted, FDEC is energized and counter CMEM is incremented. The test for @ defines the end of the message.

A flip-flop D synchronizes the start signal STA.

Bits *a, b, c, d, e, f, g* will be determined by the control unit in order to obtain the actions described above.

1b. The associated model is given in Figure 6.48.

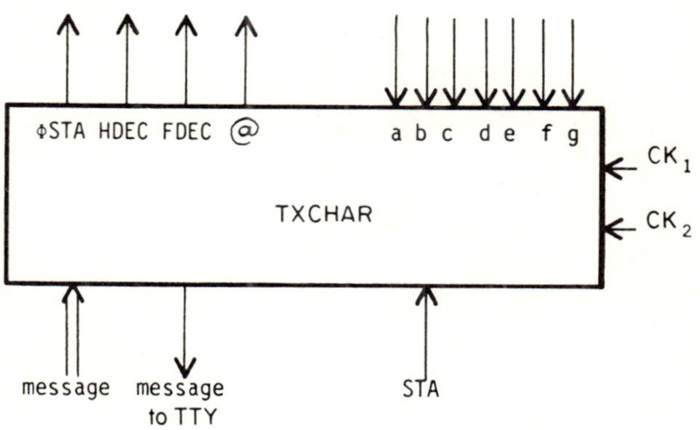

Figure 6.48

2. The state diagram is given in Figure 6.49. The numbers in circles represent the different stages of operation.

We have indicated the clock functions. We shall explain these points when studying the control unit.

3. The memory contents are given in Figure 6.50.

The corresponding memory bytes, located by their address, have been obtained from Figure 6.26.

4. The timing diagrams are given in Figure 6.51.This shows the functions described in problem 1.

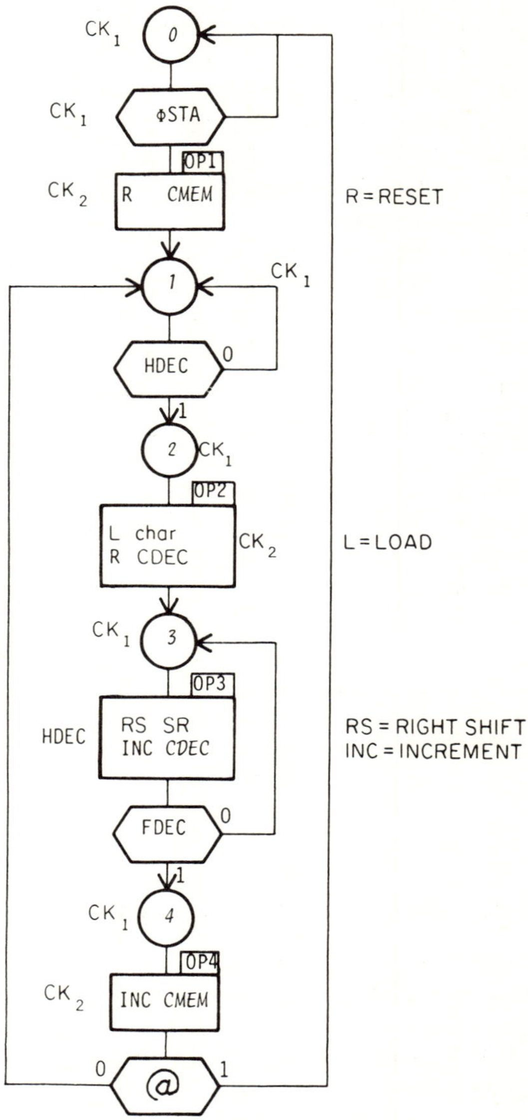
CK1
0
CK1
ΦSTA
OP1
CK2
R CMEM
R = RESET
CK1
1
HDEC
0
1
2
CK1
OP2
L char
R CDEC
CK2
L = LOAD
CK1
3
OP3
HDEC
RS SR
INC CDEC
RS = RIGHT SHIFT
INC = INCREMENT
FDEC
0
1
CK1
4
OP4
CK2
INC CMEM
0
@
1

Figure 6.49

address →	0	1	2	3	4	5	6	7	8	9	10	11	12	13	14	15
ASCII →	0A	0D	20	20	46	45	4C	49	58	20	51	55	49	20	0A	50
message →	LF	CR	SP	SP	F	E	L	I	X	SP	Q	U	I	SP	LF	P

16	17	18	19	20	21	22	23	24	25	26	27	28	29	30	31
4F	54	55	49	54	20	52	45	52	55	4D	0A	0D	20	20	20
O	T	U	I	T	SP	R	E	R	U	M	LF	CR	SP	SP	SP

32	33	34	35	36	37	38	39	40	41	42	43	44	45	46	47
20	43	4F	47	4E	4F	53	43	45	52	45	0A	08	08	43	41
SP	C	O	G	N	O	S	C	E	R	E	LF	BS	BS	C	A

48	49	50	51	52	53	54	55	56	57	58	59	60	61	62	63
55	53	41	53	0A	0A	0D	20	20	20	56	49	52	47	49	4C
U	S	A	S	LF	LF	LF	SP	SP	SP	V	I	R	G	I	L

64	65	66	67	68	69	70	71	72	73	74	75	76	77	78	79
45	20	20	28	47	45	4F	52	47	49	51	55	45	53	29	
E	SP	SP	(	G	E	O	R	G	I	Q	U	E	S	)	

LF = Line feed

CR = Carriage return

SP = Space

BS = Back space

Figure 6.50

6.S.4 Design of the control unit for the binary divider

1. The state diagram is given in Figure 6.52.

State 1 allows the execution signal EX to be awaited.

State 2 initialises the system

In state 3, a left shift brings the first bit of DVE into the accumulator and the counter is incremented.

In state 4, the divisor is compared with the contents of the accumulator. This comparison is made by subtraction, from which a partial remainder and a carry is obtained.

State 5 allows a choice of operation to be performed according to the carry bit C_8.

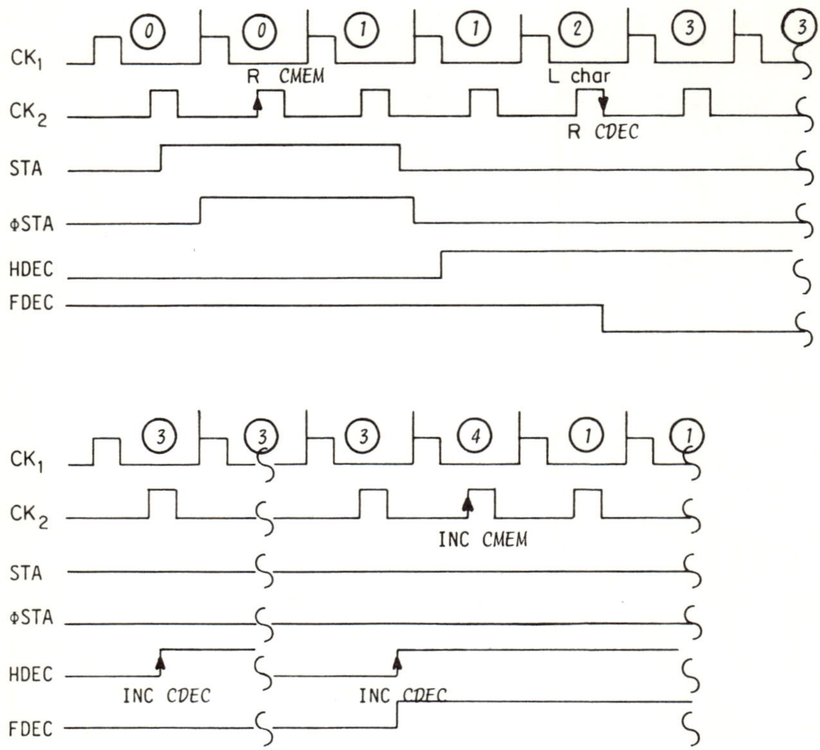

Figure 6.51

State 6 corresponds to the case where (ACCU) < (DVR); in this case the preceding partial remainder is restored by addition and the least significant bit of register Q (LSBQ) is set to 0: in fact this is the required quotient bit for this stage of operation.

State 9 corresponds to the case where (ACCU) ⩾ (DVR); in this case the partial remainder is not restored and LSBQ = 1.

After states 6 and 9, J, which indicates the end of the calculation, is tested in state 7. There is no problem in the case where 6 precedes 7 from an address point of view.

As it is not possible to go from 9 to 7, it is necessary to insert a fictitious test F, always at 0, into state 10 in order to reach state 7. In that state, J can be tested leading to state 3 or 1.

It is also necessary to put a fictitious test F in state 8 which follows 7 if J = 1.

There are ten states, which require four bits to code them.

2. The detailed diagram is given in Figure 6.53.

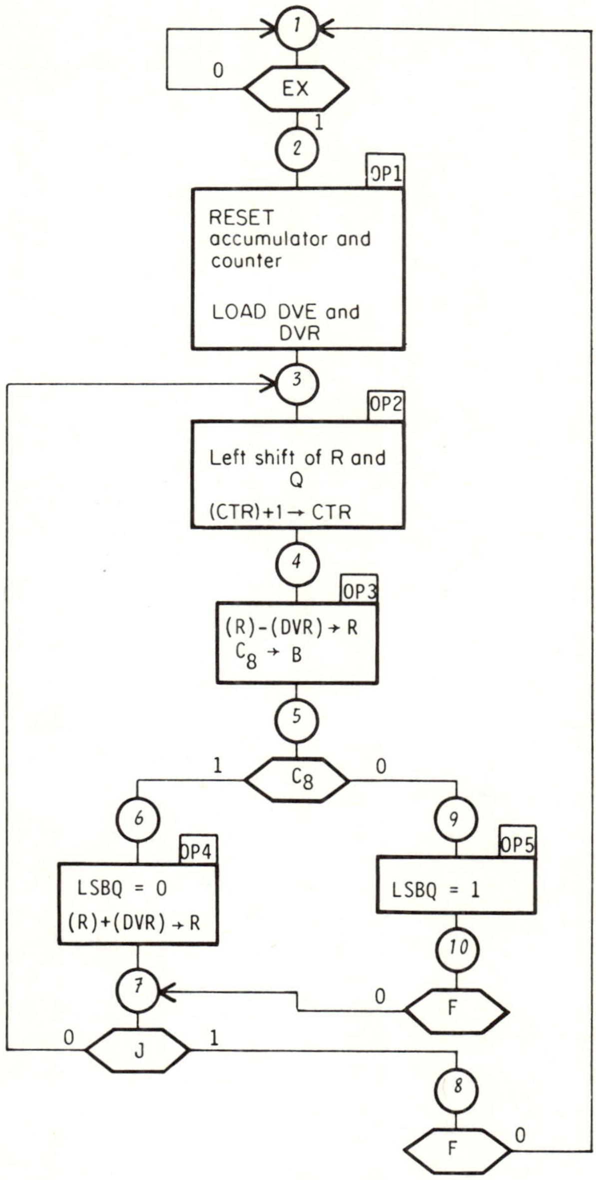

1
0
EX
1
2
OP1
RESET
accumulator and
counter
LOAD DVE and
DVR
3
OP2
Left shift of R and
Q
(CTR)+1 → CTR
4
OP3
(R)-(DVR) → R
C8 → B
5
1
C8
0
6
OP4
LSBQ = 0
(R)+(DVR) → R
9
OP5
LSBQ = 1
10
7
0
F
0
J
1
8
F
0

Figure 6.52

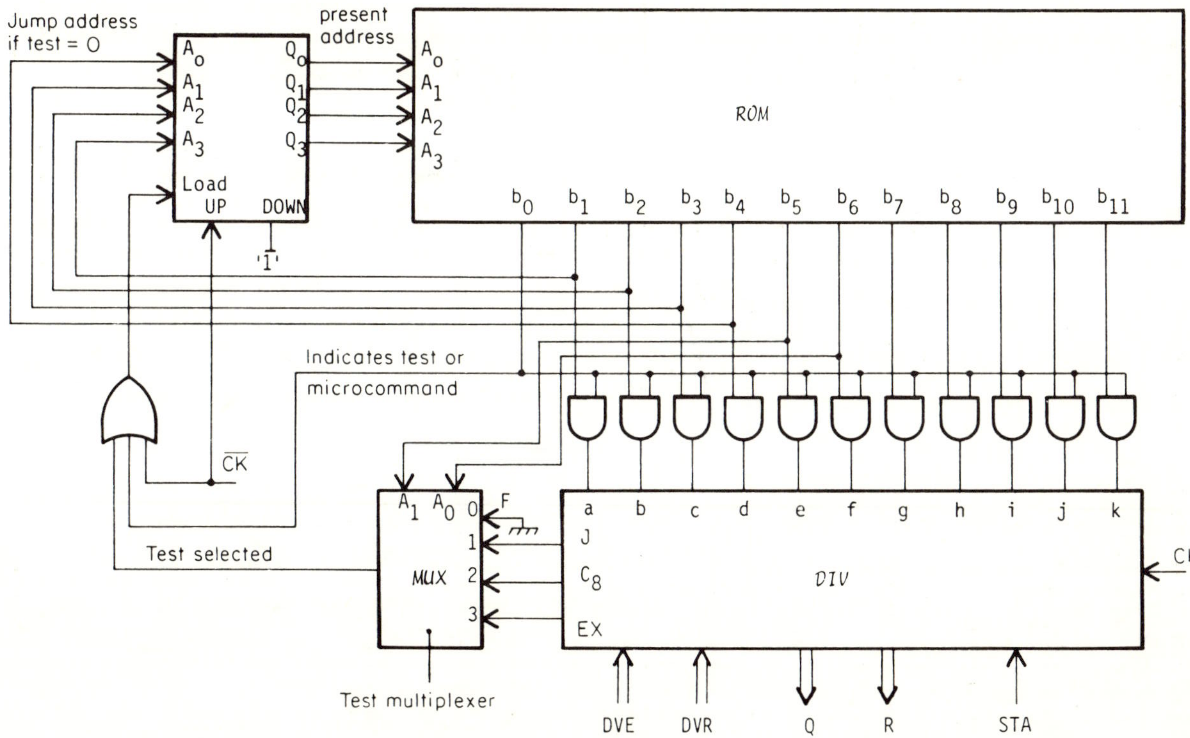
Jump address
if test = 0
present
address
A_0 A_1 A_2 A_3
Q_0 Q_1 Q_2 Q_3
Load
UP
DOWN
'1'
ROM
b_0 b_1 b_2 b_3 b_4 b_5 b_6 b_7 b_8 b_9 b_{10} b_{11}
Indicates test or
microcommand
$\overline{CK}$
Test selected
MUX
F
Test multiplexer
a b c d e f g h i j k
J
C_8
EX
DIV
CK
DVE
DVR
Q
R
STA

Figure 6.53

When $b_0 = 0$, the microcommand word does not operate, this corresponds to inhibition of the processing unit; bits b_1 to b_4 code the jump address and bits b_5 and b_6 allow selection of the associated test at the current address. If this test is at 1, the Load input of the counter is inactive, so the next state will be the present state + 1. If the test is at 0, the load input is activated by a clock pulse and the jump address is loaded.

When $b_0 = 1$, the Load input is inactive; bits b_1 to b_{11} provide the microcommand word.

Specification of the clock

The control unit and the processing unit each require a clock. Let CK_1 be the first and CK_2 the second. Figure 6.54 shows the operation of each one. We have shown the different propagation times and the corresponding available information.

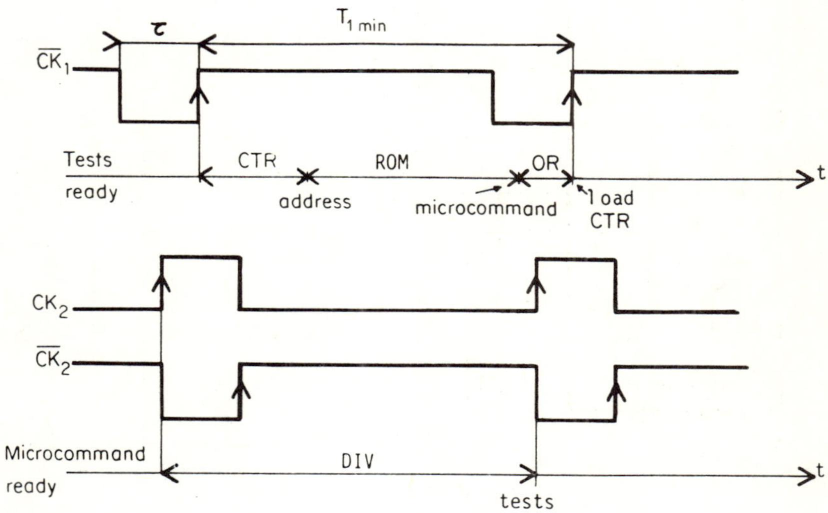

Figure 6.54

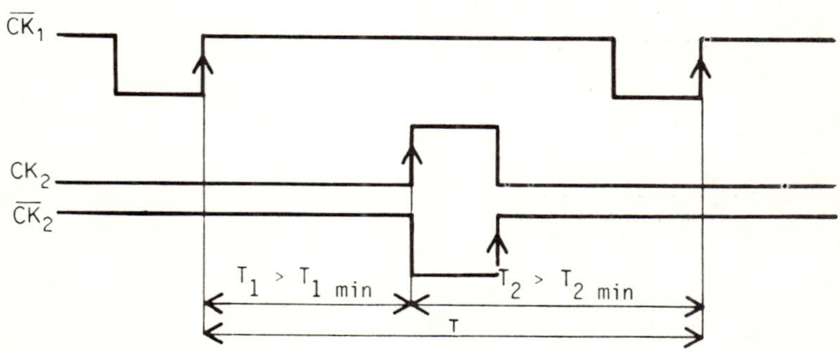

Figure 6.55

CK_1 is active on the rising edge.

CK_2 is active on the two edges separated by τ.

One solution consists of choosing two clocks of the same period with a relative phase difference, as shown in Figure 6.55. In this case $T = T_1 + T_2$, these values being determined from Figure 6.54.

It is, however, possible to use the same clock CK, as shown in Figure 6.56. We have assumed that t_{pd} OR $< \tau$, where $t_{pd}x$ is the propagation time of component x. The processing unit executes the stable microcommand while the control unit evalutes the new address from the tests which must not change.

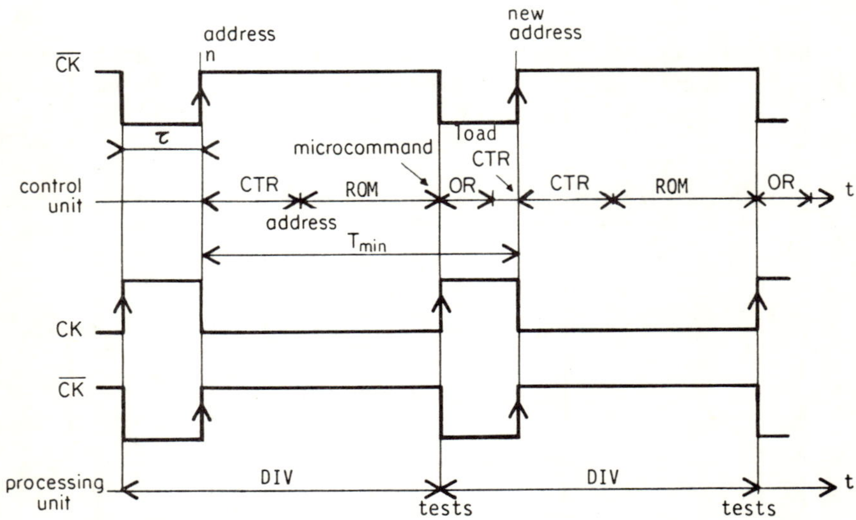

Figure 6.56

present address	b_0	ADJMP	ADTST	test selected	micro-command
1	0	1	3	DEP	
2	1				0P1
3	1				0P2
4	1				0P3
5	0	9	2	C8	
6	1				0P4
7	0	3	1	J	
8	0	1	0	F	
9	1				0P5
10	0	7	0	F	

Figure 6.57

The arrangement functions correctly if:

$$t_{pd}\mathrm{DIV} < T, \text{ where } T = (t_{pd}\mathrm{CTR} + t_{pd}\mathrm{ROM} + \tau)_{max}.$$

3. The microprogram loaded into memory is detailed in Figure 6.57.

This table, in accordance with the state diagram of Figure 6.52, shows that:

— if $b_0 = 0$, the memory word represents ADJMP and ADTST,

— if $b_0 = 1$, the memory word represents a microcommand designated by OP_i.

The detailed program is given in Figure 6.58. It is in accordance with the diagram of the processing unit in Figure 6.45. It ensures execution of the microcommands OP_i.

addr. hex.	b_0	ADJMP b_1	b_2	b_3	b_4	ADTST b_5	b_6	b_7	b_8	b_9	b_{10}	b_{11}	ADJMP-ADTST-Φ if $b_0 = 0$; OP*i* if $b_0 = 1$	
1	0	0	0	0	1	1	1	–	–	–	–	–	1	3
2	1	0	0	1	1	0	–	1	–	–	0	1		OP1
3	1	1	0	1	0	–	–	0	–	–	1	0		OP2
4	1	1	1	0	0	–	–	0	1	1	0	0		OP3
5	0	1	0	0	1	1	0	–	–	–	–	–	9	2
6	1	1	1	1	1	1	0	0	0	–	0	0		OP4
7	0	0	0	1	1	0	1	–	–	–	–	–	3	1
8	0	0	0	0	1	0	0	–	–	–	–	–	1	0
9	1	0	0	1	1	1	1	0	0	–	0	0		OP5
A	0	0	1	1	1	0	0	–	–	–	–	–	7	0
		a	*b*	*c*	*d*	*e*	*f*	*g*	*h*	*i*	*j*	*k*		

Figure 6.58

6.S.5 Design of the control unit for the character transmission system

1. The state diagram given in Figure 6.49 observes the constraints of the method.

An address includes a group of compatible actions followed by a possible test. This is the case for states 3 and 4.

State 1 includes a test without an action.

State 2 includes actions without an apparent test, a fictitious test F must be included.

State zero is exceptional: an action, performed by clock CK_2, follows a test obtained from another clock CK_1.

There are five states, which require three bits for coding.

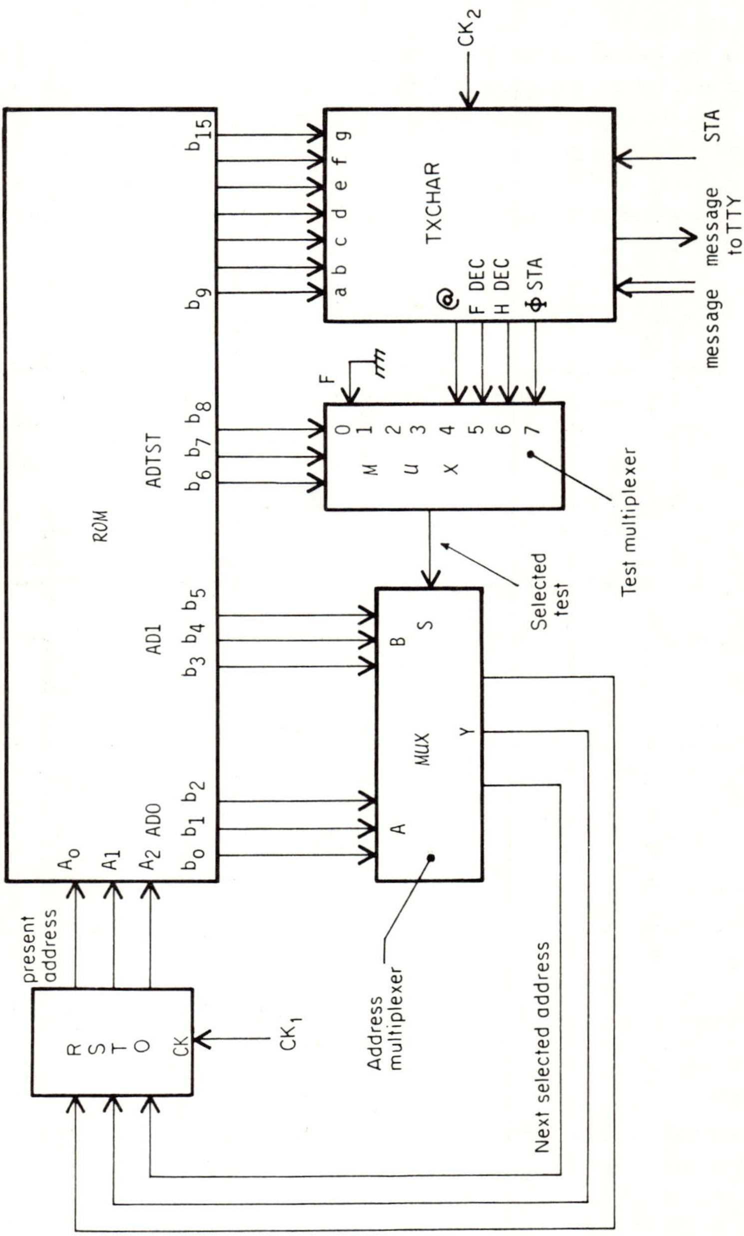

ROM
A_0
A_1
A_2
ADO
AD1
ADTST
b_0 b_1 b_2
b_3 b_4 b_5
b_6 b_7 b_8
b_9
b_{15}
TXCHAR
a b c d e f g
@
F DEC
H DEC
Φ STA
CK_2
STA
message
message to TTY
F
0 1 2 3 4 5 6 7
MUX
Test multiplexer
Selected test
A
B
S
Y
MUX
Address multiplexer
R S T O
CK
CK_1
present address
Next selected address

Figure 6.59

2. The detailed diagram is given in Figure 6.59. A test address ADTST is associated with each state-address. The selected test drives the address multiplexer:
— if test = 0, the next address is AD0,
— if test = 1, the next address is AD1.

Specification of the clocks

Figure 6.60 shows the operation of each of the clocks CK_1 and CK_2.

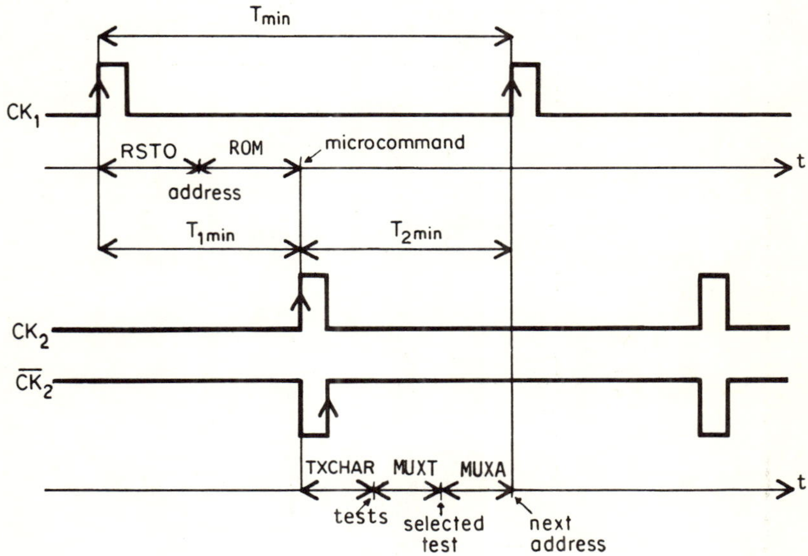

Figure 6.60

It is evident that:

$$T_1 \min = t_{pd}\mathrm{RSTO} + t_{pd}\mathrm{ROM}, \quad \text{and} \quad T_2 \min = t_{pd}\mathrm{TXCHAR} + t_{pd}\mathrm{MUXT} + t_{pd}\mathrm{MUXA}.$$

We choose $T_1 > T_1 \min$ and $T_2 > T_2 \min$, so that $T = T_1 + T_2$. A simple solution consists of taking $T/2$ = the greater of T_1 and T_2.

present address	next addresses ADD	AD1	test address ADTDT	selected test	micro-command
0	0	1	7	ϕSTA	OP1
1	1	2	6	HDEC	
2	3	—	0	F	OP2
3	3	4	5	FDEC	OP3
4	1	0	4	@	OP4

3. The microprogram to be loaded into memory is detailed in Figure 6.61. This table is in accordance with the state diagram of Figure 6.49.

The programming details are given in Figure 6.62. It is in accordance with the diagram of the processing unit in Figure 6.47. It ensures execution of the microcommands OP_i.

	AD0			AD1			ADTST			*a*	*b*	*c*	*d*	*e*	*f*	*g*				
addr	b_0	b_1	b_2	b_3	b_4	b_5	b_6	b_7	b_8	b_9	b_{10}	b_{11}	b_{12}	b_{13}	b_{14}	b_{15}	AD0	AD1	ADTST	OP*i*
0	0	0	0	0	0	1	1	1	1	1	0	0	0	0	0	0	0	1	7	OP_i
1	0	0	1	0	1	0	1	1	0	0	0	0	0	0	0	0	1	2	6	
2	0	1	1	—	—	—	0	0	0	0	0	1	1	1	0	1	3	—	0	OP2
3	0	1	1	1	0	0	1	0	1	0	0	1	0	0	1	0	3	4	5	OP3
4	0	0	1	0	0	0	1	0	0	0	1	0	0	0	0	0	1	0	4	OP4

Figure 6.62

In conclusion, we hope that this work meets the wishes expressed by many of our listeners: students and professionals in industry.

All observations will be welcome and will help in the possible production of a second edition.

Bibliography

Traite des Ordinateurs
A. Peticlerc, Dunod

Logique et Organes des Calculatrices Numeriques
G. Boulaye, Dunod

Practical Digital Design Using IC's
J. D. Greenfield, John Wiley & Sons

Methode de Synthese des Systems Digitaux
A. Stauffer, Bulletion ASE/UCS, T71 (1980) n°3 et 7

Designing Logic Systems Using State Machines
Ch. R. Clare, McGraw-Hill

Logique Combinatoire
J. Lagasse et Coll., Collection Dunod Université

Logique Sequentielle
J. Lagasse et Coll., Collection Dunod Université

Documents Motorola—Texas Instruments

Index